Foresight is 2021

Southeast Missouri State University
Historic Preservation Program
Retrospective

1981-2021

40th Anniversary Ruby Jubilee

by
David W. Jackson

The Orderly Pack Rat
2021

Jackson, David W. *Southeast Missouri State University's Historic Preservation Program Retrospective: Foresight is 2021—40th Anniversary Ruby Jubilee, 1981-2021.* (Kansas City, Mo.: The Orderly Pack Rat, 2021).

Jackson, David W., 1969-
Southeast Missouri State University's Historic Preservation Program Retrospective: Foresight is 2021—40th Anniversary Ruby Jubilee, 1981-2021.
218 p. cm.
Includes bibliographical references, illustrations, and index.

ISBN-13: 978-1-7343686-5-9 (The Orderly Pack Rat)

1. Historic Preservation—Missouri. 2. Social Science—General. 3. Science—Study and teaching. 4. Humanities—Study and teaching (Higher)—United States. 5. Learning and Scholarship. I. Jackson, David W. 1969-. II. Title.

40th Anniversary Ruby Jubilee Edition, 1981-2021, November 2021.

Cover and book design by David W. Jackson, The Orderly Pack Rat.

Cover image: Dorothy Gale's ruby red slippers conserved by Dawn Wallace, Smithsonian's National Museum of American History (americanhistory.si.edu/blog/ conservation-ruby-slippers) overlayed in front of Carnahan Hall, SEMO.

Published by:
The Orderly Pack Rat
Kansas City, Missouri
david.jackson@orderlypackrat.com
orderlypackrat.com

Dedication

to

Dr. Arthur H. Mattingly
Dr. Frank Nickell
Dr. Art Rosser
Bob White

Four visionaries whose legacy commenced in 1981
survives and thrives after 40 years.

May your hundreds of protégé,
Southeast Missouri State University
Historic Preservation Program Alumni,
carry your vision forward into
the next quadragennial . . . and beyond.

Table of Contents

The SEMO HP Alumni Chapter produces an alumni directory, offered separately to its supporting members.

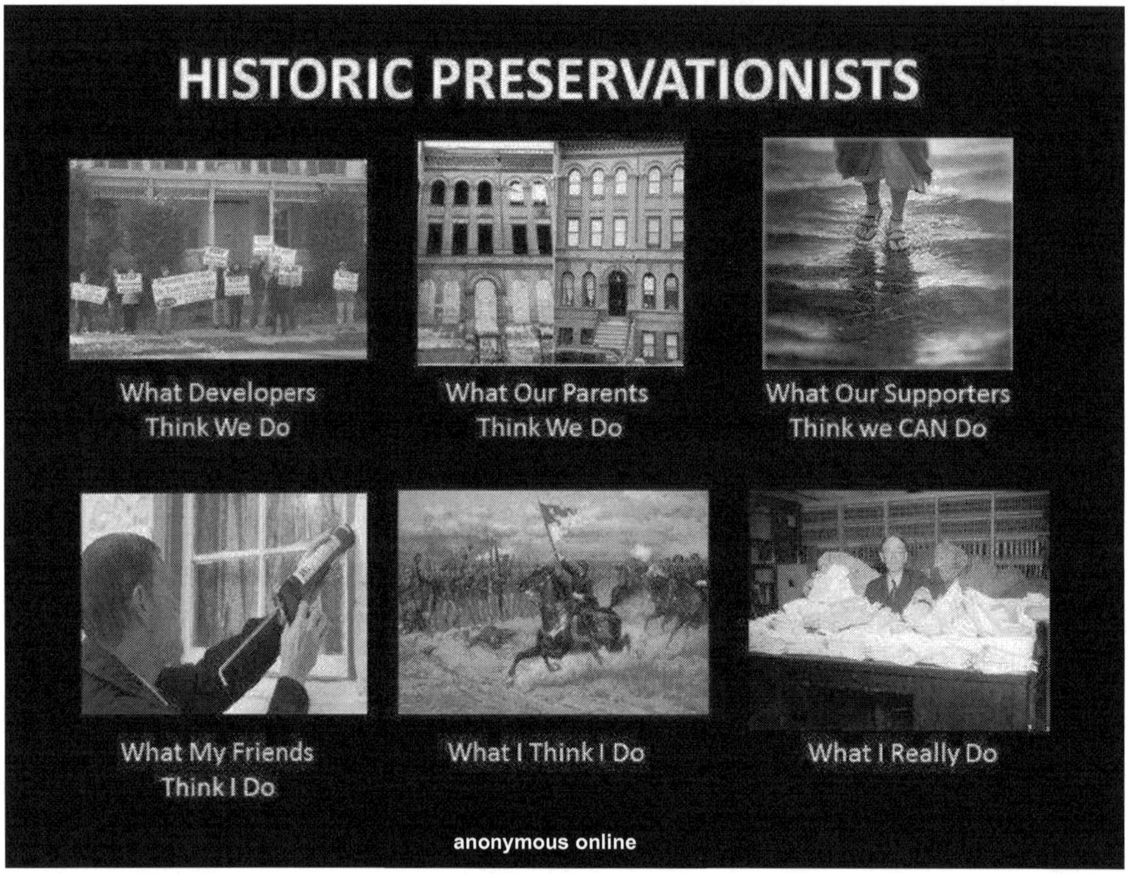
HISTORIC PRESERVATIONISTS
What Developers Think We Do
What Our Parents Think We Do
What Our Supporters Think we CAN Do
What My Friends Think I Do
What I Think I Do
What I Really Do
anonymous online

Introduction

Southeast Missouri State University's (SEMO) Historic Preservation (HP) Program—after years of planning and negotiating—was formally approved by Missouri's Coordinating Board for Higher Education (CBHE) on March 8, 1980. The degree program launched in Autumn 1981. Only four universities in America at that time boasted an undergraduate program of this kind, and SEMO was one of them.[1] Today, 40 years later in 2021, there are nearly 50 historic preservation-related degree programs.[2]

Eleven students were among the first graduating class to earn a Bachelor of Science Degree in Historic Preservation (BSHP) from SEMO in Spring 1982. Some continued to earn master's degrees, and at least five have attained their doctorate. Each have been and continue to be influential in and supportive of the field of Historic Preservation, and of SEMO's HP Program.

In Autumn 2021, SEMO's administration, faculty, students, and alumni reflect on the HP Program's 40-year track record. A cadre of pioneering Missouri educators have lectured, trained, and advanced nearly 450 students with an accredited degree into the specialized field of historic preservation.

The SEMO HP Program and its curriculum have transformed over time to prepare career professionals for an everchanging field. Three core tenants—all being "historic firsts" for SEMO—that have remained steady from the beginning, however, are that each student will have:

1) *Discovered and been drawn to SEMO's unique HP Program from hometowns across the United States;*

2) *Participated in a semester-long, degree- and area-specific internship for a minimum of nine credit hours (three times the length and strength of a traditional internship); and,*

3) *Experienced an interdisciplinary curriculum, drawing on courses and instruction from across SEMO's various departments and colleges.*

Moreover, each alum charges forth to serve in a multitude of disciplines that brings credit and prestige to their alma mater. Many forge careers "in the field" of historic preservation-related positions. Others employ their knowledge and passion no matter what path in life and profession they lead. Graduates' lifelong beliefs in preservation and conservation, together with the skills acquired at SEMO, guide them every step of their way. Some then, find ways to give back to SEMO's HP Program.

HP alumni are invited to join and take advantage of their newly developed alumni chapter, *SEMO's first-ever, degree specific alumni chapter*.

And there are specific, endowed, and unrestricted funds in which to choose to make each financial gift a lasting contribution to HP students' educational experience.

This commemorative book, the first attempt at such a compilation in nearly a half century, documents the formation, evolution, challenges, and triumphs of the SEMO HP Program. The author asks readers who discover missing or inaccurate data to understand the complexities of such a task. Then, assist in making future editions better by a) sharing documented submissions, and b) donating any *original* HP Program or Historic Preservation Association (HPA)-related documents and/or photographs from their time at SEMO to the Special Collections and Archives Department in Kent Library.

SEMO HP student Allyson Ingli worked with a small group of HP students who designed, developed, and installed an interpretive exhibit at the Cape River Heritage Museum presenting highlights of the SEMO HP Program's 40-year history. Digital images of the case and panels will be posted online into the future on the SEMO HP Alumni Chapter's website, and printed reproductions may also be archived with the University Special Collections and Archives Department in Kent Library.

Two groundbreaking announcements were published by SEMO's Department of History in Spring 1981 that would have significant impact to the history and culture of SEMO. The first was the debut of the Bachelor of Science Degree in Historic Preservation. The second was the opening of the Center for Regional History. Each are developed in the history that follows. The legacy of these fateful moments in time includes an expanded awareness of historic preservation in and around Cape Girardeau and all of Southeast Missouri.

Moreover, a regiment of hundreds of SEMO HP alumni are stationed from shore to shore. Former SEMO students—whether in an official position in the field or not—daily advocate for preservation matters. They might argue for the adaptive reuse of historic buildings . . . from grass roots and root cellars to the penthouse suites and sweet, spent houses.

Some strive to preserve two-dimensional textual paper and photographic specimens (or even audio-visual and electronic resources). Still others are adept at conserving textile resources and all sorts of three-dimensional artifacts. They all—individually and collectively—make objects of enduring historical significance safely available to the American public. They educate on history, heritage, and preservation subjects. And SEMO HP alumni strenuously encourage the creation and dissemination of cultural, educational, interpretive programs, and exhibitions.

Ultimately, innumerable vibrant historic districts, museums, archives, historical societies, and living history programs abound today (2021) because of the SEMO HP Program, which after 40 years of measured progress, has set deep, historic roots of its own at SEMO, in and around Cape Girardeau, and across the United States of America.

Lift your French 75s, give three cheers to the SEMO HP Program founders. Hip hip hooray!

And toast to the next 40 years of continued progress and success, one and all!

If "Hindsight is 2020…"
Foresight is 2021.

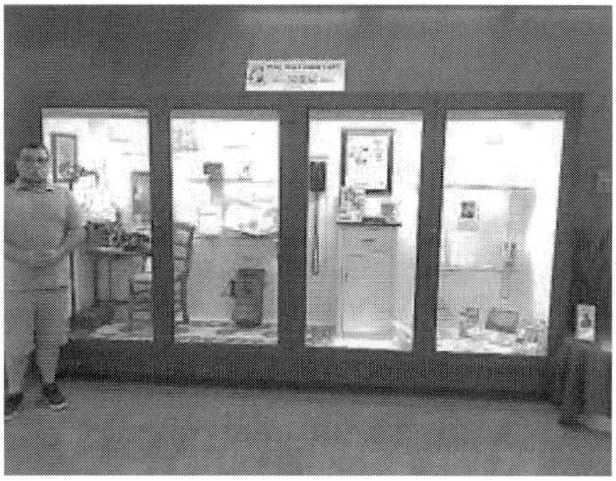

1

Origins of an Historic Curriculum

The seeds that took root and grew into a degree program in historic preservation at Southeast Missouri State University were broadcast to Cape Girardeau, Missourians, in the early 1970s. While the National Historic Preservation Act of 1966 was less than a decade old, the concept of historic preservation stemmed more than a century before when Ann Pamela Cunningham (1816-1875) founded the Mount Vernon Ladies' Association in 1853, to save the home of U.S. President General George Washington.[3] For more on this pivotal moment in American History (and preservation), readers are invited to enjoy an "HP 100" course in SEMO's HP Program.[4]

Ann Pamela Cunningham
courtesy achp.gov

Historic Preservation in America in the late 1960s was catapulting just in time as destructive "urban renewal" programs across the country were decimating historic structures, landscapes, and entire districts.

^ Mount Vernon
courtesy loc.gov
> Mount Vernon Ladies Association
mountvernon.org

Mount Vernon Record.

Devoted to the Purchase of the Home and Grave of Washington.

VOL. I. PHILADELPHIA, JULY, 1858. NO. 1.

WASHINGTON.

GREAT WASHINGTON! in virtue all sublime!
The gift of God, the heritage of time!
Oh, name immortal! of celestial birth;
Grand as the mountain monuments of earth;
Unfading as the stars that glow by night;
Refulgent as the Sun's unclouded light;
Sacred to childhood, early taught the word;
Guide of the lands by revolution stirred;
Revered by all whom heaven's broad arches span;
Graved on the heart of universal man;
Unspotted as the snow that Winter brings;
Melodious as the song that freedom sings;
Bright in the past as all that patriots claim,
Splendid in future as eternal fame.

AN APPEAL FOR MOUNT VERNON.

The following admirable appeal of "*The Mount Vernon Ladies' Association of the Union*" should go to the heart of every patriot of the land, and arouse new interest and zeal in a glorious cause. It is proper to state that the "Southern Matron," in signing her baptismal name to this document, yielded to the solicitations of Mr. Everett and other friends.

MOUNT VERNON—THE PROPERTY OF THE NATION.

It is with feelings of the highest gratification we announce to the public that the persevering efforts of the Mount Vernon Ladies' Association of the Union have been crowned with success commensurate with their lofty aims and sacred patriotism, and that they have at last secured to the American people the privilege of making the Home and Grave of WASHINGTON *the property of the Nation!*

On the failure of the Mount Vernon Bill in the General Assembly of Virginia, we renewed our overtures to Mr. Washington to sell the property to the Association. These overtures met with a favorable response, and on the 6th of April a contract was signed by Mr. Washington, which obligated him to transfer Mount Vernon to the Association, on its compliance with the following terms, viz:—

The payment of two hundred thousand dollars, for two hundred acres of land, including the mansion, gardens, landing-place, and, above all, the tomb.

Eighteen thousand dollars to be paid on closing of the contract, and the remainder of the sum in four bonds, payable in yearly instalments, with the permission, *after payment of the first bond,* to pay to Mr. Washington any amount of the balance due, in sums of not less than five thousand dollars, which sums will be credited to the Association—in this manner lessening the interest.

The title to the estate, and possession, to be given on payment of the principal and interest; and the privilege also granted of obtaining possession, on thirty days' notice, *at any time the Association may be ready to furnish the entire purchase money.*

The price asked for Mount Vernon by its proprietor, and which has never varied, has long been known to the public; and the Ladies of the Mount Vernon Association, in offering to become its purchasers, *could not expect to acquire it* upon other terms than those given as the ultimatum, to Congress and to Virginia.

Yet there is a concession made to the Association by Mr. Washington, as to the extension of time; and a very important *proviso,* to save interest, which can best be explained by the following letter, addressed to the legal gentleman who negotiated the purchase on behalf of the Association:—

Richmond, April 7, 1858.

To W. H. McFARLAND, ESQ., AND GEN. A. A. CHAPMAN:

GENTLEMEN—In the negotiation you have recently concluded with me, on behalf of the Ladies' Association for the purchase of Mount Vernon, the payment of interest on the deferred instalments from the date of the contract, while I retained possion of the property, was the only serious difficulty between us.

Having satisfied you that I could not yield this point, you finally conceded it.

As the contract is now closed, and I believe in a satisfactory way to both parties, it affords me pleasure to say to you, that if the Association meets its engagements with me on the first of January, 1859, and thinks proper to pay, and does pay the other deferred instalments on or before the 22d of February, 1859, as you inform me they will probably be both willing and able to do, it is my intention, in that case, to remit to the Association all the interest on all the deferred payments.

This I shall be willing to do, partly as an inducement to the Association to obtain possession of Mount Vernon at an early day, but more particularly as an evidence of my appreciation of the motives that have actuated the Ladies' Association in the cause it has undertaken.

I have the honor to be, gentlemen,
With the highest respect,
Your obedient servant,
JOHN A. WASHINGTON.

More than half a century has now elapsed since WASHINGTON walked among us—half a century of neglect and ingratitude to his memory. Between the broken pledges of Congress, and the apathy of his mother State, the remains of the Father of his Country have lain unhonored by the mausoleum voted in the hour of a Nation's grief, and his home and grave left to the fate of private property, and accessible to the public only through the forbearance or courtesy of its various private owners!

Ever and anon, a voice was heard to protest against republican ingratitude, for, ever and anon, the heart of this great nation seemed to throb with a remembrance of his virtues and of his services, and the priceless legacy he bequeathed. But, during the stern struggles of a young Republic, this voice was unheeded, and since America became a power on earth, it was either silenced by the din of party and sectional strife, or stifled by the murmurs of Mammon, until the generations which knew him not were in danger of forgetting, and wholly departing, from those *principles* which are the bulwark of our greatness, and those warning *counsels* which can alone maintain our Union in safety. But that Providence which has so often interposed for our rescue in the darkest days of the Republic, when the Ship of State, with no master hand at the helm, rocked to and fro on the angry waves of sectional strife and bitterness which threatened to engulf it; has stirred the heart of woman to revive, through the rescue of the *sacred* ashes of the Father of his Country, that love for his memory, slumbering, but not dead, which could be made *all-powerful* in regenerating and healing influences!

A call was made to *the women* of the South, to gather around his grave, and become the Vestals to keep alive the fires of patriotism. The motives were pure—the intentions generous—but they failed! Ye, who watch the signs of the times, know ye not wherefore? *Washington belonged not alone to the South!*

Again the call was made, and this time to *the women of the Nation.* Again it failed! and wherefore?

The title, and all the power, were to be given to a State; and Washington belonged not to one State alone!

Devoted woman would be neither baffled nor conquered; but she alone triumphs when the common homestead can be procured as a common heritage, for the estranged children of a common father, the spell of whose memory will yet have the power to reunite them around *his* hallowed sepulchre.

Unexpected success has crowned her efforts. Our country can be saved, one and indissoluble, forever—for woman has become her guardian spirit. In the sacred groves of Mount Vernon she will learn those lessons which will never leave us without a *Mary,* to train a Washington for her country's hour of need.

Orators—Statesmen—the noble brotherhood of Masons—Odd Fellows—Patriots, one and all, have come promptly to the aid of woman! And ye, women of the North and of the South, of the East and of the West, will ye not rally to the work? Will ye not vie, one with another, which will *give* most and *do* most, that we may, on the 22d of February, 1859, claim and take possession of the Home and Grave of him who loved us all, and thus make *his* birthday the birth-day also of Republican gratitude, justice, and fraternal love?

ANN PAMELA CUNNINGHAM,
Regent Mount Vernon Ladies' Association.

Mount Vernon Ladies Association newsletter, Vol., No. 1, July 1858
common-place-archives.org

Cape Girardeau may have fared better comparatively than other communities because of the action and advocacy of a purposeful awareness campaign waged by the Historical Association of Greater Cape Girardeau, led by its two-term president Dr. Arthur H. Mattingly. Another local organization, Old Town Cape, Inc., is a community-driven nonprofit seeking to preserve and revitalize "the historic, cultural, and economic landscapes" of its downtown sector.[5]

Mattingly submitted his 1973 "President's Report" report, which was published in the *Southeast Missourian* newspaper:[6]

The Strength of Opinion in a Living Community

"The question is often asked, 'Why preserve this house, or block, or section of the community?' I would like to devote my report this year to a consideration of a historic district for Spanish Street.

"Preservation of historic houses or blocks is a practical one. Preservation recycles the structures that are there. Restoration is not rehabilitation. Rehabilitation is renovating the structure to make it fit for habitation once again. Restoration goes beyond that to include retaining the original exterior architectural character and as much of the original interior as possible. It adds dimension of special value beyond that of structural soundness. It does not require taking down the existing structure and building it up again. It is faster and creates less wear and tear on the neighborhood.

"Preserving buildings that have architectural value please the eye and engage the mind. As our cities grow more dehumanized, as our suburban dwellings become more redundant, as our supply of old buildings dwindles, we find the older architectural styles becoming more and more important. Walking the streets of an historic district becomes a significant experience for thousands of people. People do not necessarily long to live in the past, but they need a mixture of the past and present, a reminder of the way things were and an escape from the less attractive aspects of our present cities.

"Each society as it occurs in the long course of time removes or contributes to the sum total of shelter it provides for its many functions. These shelters, singly or in groups, for homes, for work, play, worship, government, or any other purpose, in the total context of the community provide a portion of the evidence of the endurance or the lack of continuity of a society. I suggest our community assume the responsibility for placing value on the work of its past generations. This is in part to see where society has been and to judge whether it is improving itself, and to develop a firm tradition which will enable future societies to identify all of us in the twentieth century to the generations of the future.

"Since we are concerned with only one small part of history, that of place and objects related to place, ours is not too complex a task. Only a small part of historic preservation is mathematical.

Most of it is subjective, a matter of opinion. For instance, do we tear down the Common Pleas Courthouse or don't we? Is the value of the land on which it sits greater than the value of the building's historic and architectural merit? Is the fact that there is no immediate and practical use for the building important? We may be fond of the building. It means something emotionally to us, and we wouldn't dream of destroying the structure. Of course, the example might be fine architecture. Its history can be documented. In the opinion of many it is historic and valuable. Therefore, opinion becomes the criterion for preservation, and opinion is the food for political and public action.

"The problem of applying the strength of opinion to more than one building in a variety of areas in different places is more complex. It is more difficult to conceive of building groups as symbols when the individual structures are in themselves not always indefinable of substantial historic or architectural interest, although together they provide a community's character which symbolizes a place, era, or an architecture.

"My point is that historic preservation is not and cannot be an exact science. The word "historic" is used here to imply both history and architecture of the place. 'Historic district,' as a term, implies an area in which historic events, historic buildings, buildings of architectural value and buildings which in groups from historic associations are found, in whatever combination. I would add an important, often overwhelming, consideration in that the area or district is, or can be made, pleasing to the eye. Its location, landscape, and totality as an architectural entity, whether in one style or many, is all important. A[n] historic district is a combination of a great variety of parts, no two alike, constituting an individual place of special quality warranting perpetual care as part of the inbuilt, permanent, and irreplaceable values of a community.

"An early nineteenth century textile mill complex may be as worthy of preservation as a complete village green with its church, meeting house, and surrounding homes. The decisions must be local and based on what is to be found of historic value. The area subject to study and investigation is part of a community and is an inseparable part of its daily life. Plans and programs for the city, town, or village are a part of a potential historic district and must be worked out together. Unless we are converting our historic district into a museum like Mystic or Sturbridge Village, it has to be a vital part of the living community.

"I believe the time has come for the community to recognize its historic assets. One final and important reason for historic preservation and the establishment of a[n] historic district: tourism. It has been said, 'Everywhere you put a plaque twelve Americans will appear in the first half hour to read it,' referring to the insatiable interest in history that Americans have developed. In every state, tourism is one of the three largest revenue producers. The Thomas Edison Birthplace in Milan, Ohio (population 1,400) brings over 25,000 visitors to town annually. The general area has fifty motels. In a recent survey, the Ohio Development Department concluded that a community attracting 36,500 visitors a year could expect to receive an additional $777,000 in personal income through 111 new jobs, $144,300 in bank deposits, and $1,119,908 in added retail sales. These are but two examples. These figures can be multiplied yet higher for historic districts because they tend to draw an even wider visitation than many single sites.

"As a community we need to develop our historic houses and district. Not only will we preserve for the future, but it will give Cape Girardeau a new industry."

Mattingly's Mission

Mattingly's activism in historic preservation was not by happenstance. Rather, it was an intentional and steady fire that continued to grow within him since childhood. Mattingly kindly shared his recollections—much of which a typical HP alum will relate—about "underlying life experiences that led me down the path toward historic preservation education."

Art Mattingly at the lectern, late 1960s

"I have always had an interest in history. My mother was a schoolteacher. In 1945, when I was [about] ten, she attended summer school in Washington, D.C. I went with her because I had no one to stay with while she was gone. We stayed out in Falls Church, Virginia, where I would ride a city bus into the District by myself and get off at 11th & E directly across from Ford's Theatre.

"I would then spend the afternoon absorbing the museums on the Mall, meet my mother, and ride back to Virginia. I did this routine countless times that summer. In high school I was fortunate to have a history teacher, not a full-time coach, part-time social study teacher. She built a smoldering fire for history that I left [with when I graduated] high school. Shortly after high school, I joined the service during the latter stages of the Korean War.

"After almost four years, I entered SEMO and graduated with a Social Studies degree. Then with what was left of the G.I. Bill education money, I entered the University of Missouri and received a master's degree in American History. [I] taught history at North

Kansas City high school. In 1964, I was offered an instructor's position at SEMO teaching survey classes of American History. In 1968, I was approved for a leave of absence to pursue doctoral studies [and] return[ed] to the campus in 1971 to complete the dissertation. . . . In 1972, I was a newly minted Ph.D. in the SEMO Department of History working again with some of my former professors from my undergraduate training.

"Through all of the above I kept my love of and for history, especially local, and state history. [I kept] busy joining the Association for State and Local History, traveling to their national meetings, and going to seminars on various topics related to local and state history.

"I offered the first graduate seminar in state and local history in the Department of History and assumed the teaching of Missouri history when Professor Brown died. Throughout this [period] at SEMO, I had taken the opportunity to visit museums, historic sites, and used archival resources at the local, state, and national repositories in my teaching and dissertation research.

"In the mid-70's I pushed for microfilm copies of Missouri newspapers to be located at Kent Library (instead of [only] at the State Historical Society of Missouri in Columbia) in order to facilitate undergraduate and graduate research by our students. This was eventually accomplished, [and today] we now have a [regional repository] of the State Historical Society located on the edge of SEMO's campus!

"In this same time period, Dr. Frank Nickell, Dr. Paul Travis, and I received a small (very small) grant to travel to some Mid-west colleges/universities to look at their archives, academic programs, and outreach programs that might serve the region.

"Nickell, Travis, and I proposed the creation of the Regional History Center as a part of the Department of History. We envisioned the Center offering outreach to organizations, and to our students who were interested in local history as a field of study. The Center was approved, and I was appointed director and served in that capacity until the HP degree program was established.

"While director of the Center, I came to realize (from earlier experiences, recent visits to museums, historic sites, and archives) many historical organizations had individuals, dedicated though they might be, did not have the professional training to support the preservation of local history in its most narrow sense.

"But I also though there were a number of people who liked . . . loved . . . history but did not want to teach the subject in a formal classroom.

"If the Regional History Center was established to help, assist, and coordinate with historical organizations (informal teaching the local residents overseeing these history sites), perhaps there was a need to create a class or two with professional instruction in museum care or archives?

"This idea was discussed informally between Nickell, Travis, and I. . . . I suggested, perhaps, rather than a class or two maybe we should think in terms of a degree. We agreed I should outline my thoughts for such a program on paper for us to consider.

"The rest is history."[7]

Degree in Historic Preservation Approved

Mattingly's encouragement, education . . . and prodding . . . organically—and perhaps with sufficient fertilizer—eventually germinated. Within six years of his 1973 "President's Report," Mattingly had mustered a handful of allies in pursuit of establishing a new degree program at SEMO . . . a new program in higher education in Missouri.

Dr. Mattingly, on the 25th anniversary of the SEMO HP Program, provided a retrospective on the SMEO HP Program's origins. His comments, first published in the 2006 edition of the Department of History's newsletter, *Pastscript*, are supplemented below with additional sources as well as personal reflections conveyed during a November 5, 2010, oral history interview between Dr. Mattingly and two HP students (*set in italics*):[8]

"...I am a traditionally trained historian. I have...an undergraduate degree from here at Southeast Missouri State and I got a master's degree from the University of Missouri history department and at that day and time you didn't specialize in a master's program, you just got a general background. I left the master's program there and went to Kansas City, Missouri, and taught high school.... I taught there for a while and at the time, there was tremendous enrollment and universities couldn't get a lot of staff to do it. So, I was asked to come back to this school [SEMO] *and teach for my master's degree. So, I came back and taught for several years and went back for my Ph.D. except I had one in the field of military history....*

But I was always interested in local history. I was interested in [more] *things than just the history. I was interested in buildings, and I was interested in things like that, so I came here and taught bringing back with me this interest in things* [other] *than just traditional history.*

I am sometimes asked, what is historic preservation? It's a long story and one that I am not sure bears repeating here. Perhaps is better to ask, "How did Southeast Missouri State University come to have such an academic program?"

I always felt, as a teacher...because when I started teaching here American history was required in every department. You had to take five hours of American history. It did not matter what degree you were pursuing. It's what you had to do. There were a lot of students I came in contact with that liked history, but they didn't want to go into the classroom and teach. Well, that was something that kind of germinated in my mind and I thought for a number of years. And I became personally interested in oral history and I liked to get out and interview people and talk to them and find out what they did with their life and why they made the choice that they made and what events in their lives caused them to make changes. I came to the conclusion that there were people out there who would pursue history classes if we offered them something other than the traditional avenue. And at that time, if you were a history person and if you were interested in history, you would either get a B.A. or you had to go into social studies. There really wasn't much else. In addition to that, I have always liked to travel, and I have always liked to go to various places: museums, historic sites.... And one of the things that occurred to me as I traveled were that there were people in these places that, at these museums, at these historic sites, that were well-meaning people, but they didn't have a good background in history. And they might know this pen that Aunt Lucy used to write whatever, but in terms of the events of that community or the events of the world, they couldn't bring those together and provide what I consider an education. Teaching in a different sense. So, these things kind of came together and out of that came, um, the nutty idea that what we should do is offer some classes, perhaps even a degree that would offer individuals interested in history that didn't want to teach. There is nothing wrong with the formal teaching, but who might want to pursue a degree that would offer the opportunity to teach in a different format. Like in a museum, like at a historic site, like serving as director of a downtown district. Those are all, in my view, the way to teach and the way to bring people into their past. So, with that kind of idea germinating and bubbling up from time to time, that's how I came to it.

Dr. Frank Nickell

Prior to the inception of [SEMO's HP] Program, three history faculty members—Paul Travis, Frank Nickell, and I—had discussed from time to time the idea of broadening the appeal of traditional history course offerings with the hope of bringing more students to our department.

We talked about one or two courses and how they might apply to traditional curriculum, but nothing much came of the talk until the summer of 1979, when we applied for a university car and travel funds to visit several mid-South universities that had courses or non-traditional history programs. Some of the programs at that time, for the lack of a better name, were referred to as public history courses or programs. That summer, I think we visited three or four campuses and faculty that had had experience with the type of programs/courses we were interested in. Our visits were supposed to help us decide what we would like to pursue at Southeast; but, if anything, the visits created uncertainty about our direction. Each school's program/courses differed in approach, organization, and how they were academically administered. We wanted to focus on the cultural heritage of the nation, state and local entities.

Local history was the underlying theme in our search for an academic vehicle to carry our courses and program.

The establishment of the Center for Regional History in 1979 [sic.] fulfilled the immediate need to assist local communities in the preservation of their heritage.[9]

But it was our goal to move from local history/cultural resources of the region to create an academic program that would train individuals interested in history for something other than the traditional occupational path of social studies teaching.

In the fall of 1979, we put together a proposal for an undergraduate degree in historic preservation. At the time of our proposal there were only three such programs, one each in Maryland, Rhode Island, and Tennessee. The first degree program, at Columbia University, was a graduate program.

Art Mattingly chalking it up, late 1960s

Now, there were programs that offered you the option to study architecture for example in a historic perspective, but that was a graduate program, and all of those programs were housed in architecture schools. The three programs, um, and I think there were three, there may have been two but three is a better number. Those three programs were not housed in history departments; they were housed in a design program for example. I don't know what the other two were. My thinking was, and the people that ultimately came on board with this, my thinking was if you were going to do historic preservation or public history, you needed to have a grounding in history. You need to be able to do historic research, you need to be able to so these things.

We felt ours was a unique avenue. That we would have it housed in the history department if we could get them to go along with it, and that it would be based on historical research, and we would try to provide to those students who came to us the opportunity to work in a variety of fields that traditionally would not be educated.

Our degree proposal was unique for [Southeast Missouri State] University at the time. We insisted on an interdisciplinary program that would draw on other departments from across the campus [a first for Southeast]. We proposed courses that were non-traditional history to be taught by history faculty. We required a semester internship by each student before graduation [another first for Southeast]. We required that faculty visit each intern during the term of their internship. While other departments would be involved, the administration of the program would be in the hands of the Department of History.

These ideas and others did not meet with an immediate, enthusiastic reception.

The first hurdle for the proposed degree was the Department. In the 1970s, the curriculum was very traditional. It served as an excellent training for social studies teachers; but the proposal was not for that type of training.

We had a lot, a lot of challenges. And that is probably a diplomatic way of putting it. "They looked at you as if you were nuts."[10]

From the very beginning those of us, Mr. Bob White, Dr. Frank Nichol, the dean of at the time applied arts, Art Rosser. Those of us sat around and it was our view that this program that we were working on should be an interdisciplinary program. That while it was housed in the history department it would call on other disciplines to provide training for people who were going out to work, for example, in a museum.

Obviously, you need a history background, but you may also need interior design, a course that is in that general area so that you have some background in that area. The same is true of English. We would have the English department ultimately write a program, of course, that was for our degree program.

So, initially we believed that it needed to be an interdisciplinary degree program. When it came to the history department there were a lot of classes that the history department- at that time it was a traditional department, I mean it was very traditional- to try to get a class, for example, of military history was a struggle.

So, to suggest to our collogues that we needed a class that focused on local history was a real challenge. And I'm not going into all of the issues, but it was a real struggle. And we had a struggle within the history department, it was a struggle to get its approval, it was a struggle to get it out of the university community to approve, and once we won that battle, it was a struggle to get the university to advance on the state and ask for approval. Oh my, yes, a lot of hoops. I don't want to hurt anybody or get anyone upset, but it was very difficult.

The internal give-and-take over the proposed degree was, at times, "interesting," and enlightening! Successfully securing Department approval was only the first University step.

The proposal went campus-wide for comments, and after the required time we had to defend it before the University community.

Once this hurdle was passed, the degree proposal had to be approved by the Board of Regents.

With Board approval, the proposal left the campus and went to the Coordinating Board of Higher Education (CBHE), which had final approval for new degree programs at Missouri universities.[11]

> The Coordinating Board for Higher Education was authorized by an amendment to the Missouri Constitution in 1972 and established by statute in the Omnibus State Reorganization Act of 1974. The nine board members, one from each congressional district and a member at large, are appointed by the governor and confirmed by the Senate. The term of appointment is six years. No more than five of the nine members may be affiliated with the same political party, and members serve without compensation.

They said how many of these programs are there in the country? See? How many? Well, there are not that many, does that mean we should not have it? Because there are not that many? We think there is a need. People who work in these fields need this kind of training.

After several weeks on pins and needles, we received word from the CBHE that our proposal had been rejected. . . . *They turned us down flat. Said no way.*

The Board objected because they could not find much information about academic programs in historic preservation. The CBHE did not believe there was a market for it, and Missouri did not have anything like it.

To say those of us involved from the beginning were "upset" is a very mild expression of our real feelings.

The Board did say in their rejection letter that we could come to Jefferson City and appeal their decision. . . . *We asked permission to come back and ask that we could make a formal presentation. And that is what we did.* Drs. George Ketcham, Frank Nickell, Art Rosser, and I got an appointment with the chairman of the CBHE and drove to Jefferson City to make our case. *And I will never forget sitting in this man's office, the chair of the Department of Higher Education, sitting in his chair, legs crossed, and he was not very receptive to us at all. We made it, we said, "look, give us a chance, if we don't turn out a sufficient number of graduates over time then okay, we would understand if you closed the degree program down, but there was a need."*

[We were then placed on the CBHE's agenda under action "Item 13. New Programs: Bachelor of Science in Historic Preservation." The meeting was held at 9 o'clock in the morning on Saturday, March 8, 1980, in the Upper Classrooms of the Convocation Center of Rockhurst College in Kansas City, Missouri.]

The four of us used our charm and winning personalities, leaving the meeting with approval of the program . . . .

[Dr. Atkins commented that the . . . proposal was the most fully developed, well-presented proposal that she has had the pleasure of working with, therefore she feels quite comfortable in recommending approval of this program. Ms. Windom moved that the

Board approve the proposed program. Mr. Sprong seconded the motion, and it was duly passed.][12]

The irony is that the CBHE used our proposal as a model to follow for future degree submissions to the state.

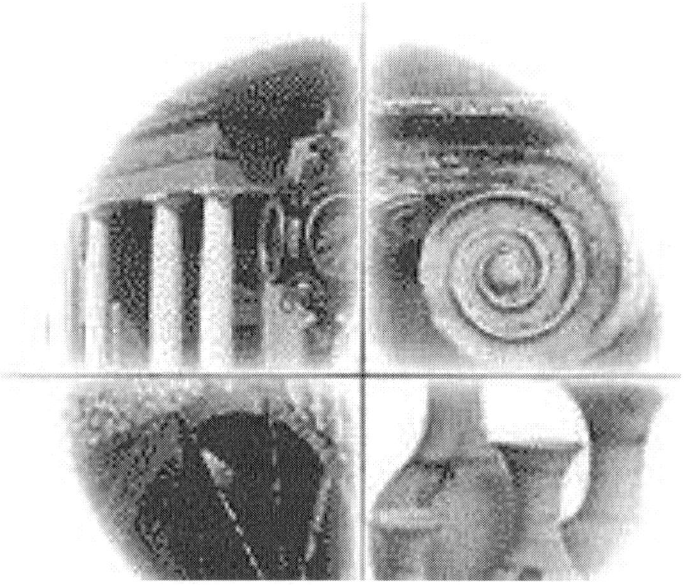

Tours of Academic Hall's dome are always special. These were taken during the 2006 Alumni Weekend.

Here are copies of SEMO HP Program's origin documents:

COORDINATING BOARD FOR HIGHER EDUCATION

Time: 9 A.M.
Saturday
March 8, 1980

Place: Upper Classrooms
Convocation Center
Rockhurst College
Kansas City

AGENDA

		Tab	Action Item
I.	Minutes of Previous Meeting	A	*
II.	Report of the Commissioner		
III.	Report of the State Librarian		
IV.	Status of FY 1981 Appropriations	B	
V.	Report of Activities of Computer Facilitating Committee	C	
VI.	Legislative Status Report	D	
VII.	Recommended Changes in the Regulations Governing the Missouri Guaranteed Student Loan Program	E	**
VIII.	Determining the Eligibility of Institutional Applicants for the Missouri Student Grant Program	F	*
IX.	Transfer Articulation	G	*
X.	Request for Modification of Program Review Procedures Relevant to New Teacher Certification Requirements	H	**
XI.	Report on Nursing Hearings	I	
XII.	Mailing of General Board Memoranda to Members of the Advisory Committee	J	*
XIII.	New Programs		
	A. Bachelor of Science in Historic Preservation Southeast Missouri State University	K	*
XIV.	Informational Items		
	A. Program Changes	L	
	B. State Library Service to the General Assembly	M	
	C. Division of Student Financial Aid Report	N	
	D. Report on the Regional Coordinating Councils for Off-Campus Education	O	

* Final Action
** Preliminary Action

MISSOURI DEPARTMENT
OF HIGHER EDUCATION
600 MONROE
JEFFERSON CITY, MISSOURI 65101
314/751-2361

BRUCE ROBERTSON
COMMISSIONER

TO: Coordinating Board for Higher Education

FROM: Albert Lynd A APPROVED BY: Bruce Robertson BR
Linda A. Atkins LA

DATE: February 29, 1980

SUBJECT: New Program Proposal
Bachelor of Science in Historic Preservation
Southeast Missouri State University
Cape Girardeau, Missouri

Southeast Missouri State University seeks Coordinating Board approval for a Bachelor of Science degree with a major in historic preservation. The proposed program has been developed within the context of Southeast Missouri State University's mission of regional service, for Southeast Missouri is an old region of the state, rich in local history and with a broad array of historical material, including archeological sites, historic buildings, historical artifacts, and manuscript notes.

Historic preservation is an outgrowth of the local history portion of a history major. As a program, it encompasses much more than history. It includes a variety of new activities, courses, and disciplines, and is a true interdisciplinary major. It includes historic site administration, museum work, archival work, and restoration of buildings and artifacts. It draws from patriotic, aesthetic, and social motivations. It has been a successful popular movement, and preservation efforts have expanded into collecting regional and national memorabilia, the restoration of historic sites, and preservation of the national landscape, to mention but a few.

As a profession, historic preservation affords diverse employment opportunities. Jobs exist with the National Park Service, State Departments of Natural Resources, and at historic sites across the state and nation. All require specialists in areas ranging from administration to real estate. Diversity frequently characterizes the jobs. For example, a preservationist, as contrasted to a museum curator, may engage in site maintenance, tour guiding, and furniture repairing. This has led to difficulties in defining the historic preservation curriculum. In the absence of a clear definable curriculum of preservation, experts often use the phrase "historic preservation core." No organized attempt has been made to spell out the content of such a "core" but there appears to be general agreement as to the goals of such educational programs. These include developing an understanding of American cultural/history

Coordinating Board for Higher Education
February 29, 1980
Page 2

and the evolution of the American environment (both man-made and natural), as well as acquiring knowledge about specific techniques that contribute to the process of preserving and protecting this environment.

Degree programs are generally defined as being one of two types: (1) occupational programs which offer a degree in historic preservation-BS/MS in historic preservation, commonly referred to as a "generalist program;" and (2) those which offer a degree in more traditional disciplines-BS/MS in history, architecture, planning, with a major or specialization in historic preservation, often called a "specialist program."

The content of a preservation program in terms of the courses offered is tied to many factors. Among the most important of these factors, as identified by the National Trust for Historic Preservation, are the purposes and focus of the sponsoring department, the backgrounds and experiences of the faculty, courses available in other departments, and the geographic location of the institution.

Another issue is the appropriateness of degree level. Just as two schools of thought guide program development--generalist versus specialist--there are two views as to the appropriate level of study. One group argues that historic preservation as a specialized field of study should only exist at the graduate level. Students should be well-versed in a traditional baccalaureate discipline such as history or architecture and then develop expertise in historic preservation by study at the graduate level. Those who suggest that historic preservation is appropriate to the undergraduate level usually subscribe to the generalists' view that the baccalaureate degree is sufficient to present a variety of career options. Students, upon completion of the baccalaureate degree are prepared to enter graduate training with more highly developed skills and interest in historic preservation than are students with an undergraduate degree in another discipline. In addition, students are trained for direct entry into historic preservation occupations.

Degree programs began in 1966. Until 1973 they were confined to the graduate level. Presently, 18 institutions offer graduate programs in historic preservation. Only three undergraduate programs exist in the nation: Middle Tennessee State University, begun in 1973; Roger Williams College, begun in 1976; and Goucher College, begun in 1976. While there are excellent arguments for and against historic preservation as an undergraduate field of study, all persons contacted by the Missouri Department of Higher Education staff agreed that tradition alone should not guide future program development. Rather, multilevel education provided by graduate and undergraduate degree programs merely reflect the uniqueness of the location, the expertise of the faculty, the short and long-term benefits for the students and the region, and the variety in the discipline.

Southeast's program is designed to meet these objectives: it will provide training for employment in the fields of historic preservation; it will prepare the student for graduate study; and it will be consistent with Southeast Missouri State University's long-range plans for optimum use of existing resources. In addition: (a) it is interdisciplinary in concept and organization; (b) it requires a field experience; and (c) it has an outreach function. The Missouri Department of Higher Education staff believes the character of the preservation profession demands an academic program that brings the student into contact with several disciplines. Moreover, the MDHE staff believes the Southeast program is not just a haphazard collection of courses; rather it brings together areas of study in a manner that is a genuine interdisciplinary learning experience.

The proposed curriculum will introduce the student to several types of experiential learning. Each student will devote at least two summers to field experiences. The field terms, as proposed, are important because they will enable the student to see, early in the program, how the program prepares him for the profession. According to the proposal, internships have already been offered by the Missouri Department of Natural Resources, the Tennessee Valley Authority, and the National Park Service.

The proposed program will enable the university to continue and expand its service role in a new way. The outreach program developing from this degree has exceptional potential not only for the local region, but for the entire state. Faculty and students will assist local, regional, and state organizations or individuals in preserving the heritages and cultures of the region.

Southeast Missouri State University, by establishing two historic preservation services: a Regional History Center and an Oral History Program, has already exploited its unique geographic opportunities in ways that will reinforce the preservation program. Furthermore, a special resource for the program is the Kent Library Regional History Collection located on the university campus. Composed primarily of historical records and materials, the collection reflects the cultural and historical contributions unique to the region. The excellent geographic location for providing unusual opportunities is repeatedly used to augment the academic program. Because the element of locational opportunity is central to this course of study, the MDHE staff believes degree programs in historic preservation should not be envisioned for all colleges and universities in the state.

The major component of the Southeast curriculum contains the historic preservation core, the interdisciplinary courses, course work in the option selected, and the field experience in the option. The requirements for the proposed curriculum are: general education--45 hours;

Coordinating Board for Higher Education
February 29, 1980
Page 4

historic preservation (three courses)--nine hours; interdisciplinary courses (from six disciplines)--39-40 hours; option (chosen from archives, museum, historic sites, and historic restoration)--12-15 hours; and electives--25-24 hours; for a total graduation requirement of 124 hours.

In most instances, existing courses will be utilized, although several new support courses will be necessary. These courses will be staffed by reassignment of present faculty, and the need for additional faculty is not anticipated. The present faculty seems well qualified in historic preservation as well as the traditional disciplines. Library and research resources are adequate to support the program. Facilities and equipment necessary to initiate the program are presently available and no new space or major items of equipment will be needed.

The program will be funded through a portion of student fees and state general revenue appropriations. The total cost of this program should not impact significantly on the institution's budget. Administration of the proposed program will be in the Department of History in the College of Social Sciences. To provide for the interdepartmental character of the program, there will be an advisory committee chaired by the program director and including representatives of the departments contributing to the program.

RECOMMENDATION: It is recommended that the Coordinating Board for Higher Educaton approve the Bachelor of Science degree in historic preservation, submitted by Southeast Missouri State University.

AL:fht

REVIEW BY CRITERIA

Bachelor of Science in Historic Preservation

Southeast Missouri State University

Role and Scope of Institution

As a regional multipurpose institution, Southeast Missouri State University includes in its objectives career preparation of students in selected areas, development of research skills, the preparation of students for graduate study, and service to its geographic region. Officials of Southeast Missouri State University believe the proposed program in historic preservation meets all of these objectives.

The proposed course of study is designed to provide training necessary for direct entry into the field of historic site administration. Furthermore, the program includes an emphasis on research techniques and is constructed to lead students into graduate training, as well as into employment. According to the proposal, the program, in conjunction with the university's Regional History Center, will provide an important service to the geographic region by serving as a resource center for persons interested in historic preservation.

In addition, this course of study has been developed within a context of long-range planning which promotes new programs for attracting students, and which stresses innovative approaches to the use of existing resources.

With these rationale in mind, the Missouri Department of Higher Education staff believes a course of study in historic preservation is consistent with the role and scope of Southeast Missouri State University.

Program Structure

In the absence of an official accrediting agency's guidelines for curriculum or program structure, the MDHE staff reviewed entries in the National Trust for Historic Preservation publicaton, Guide to Degree Programs in Historic Preservation. This review revealed a wide range and variety of academic offerings. In most degree programs, a core of required courses leads to a degree in historic preservation or to a degree in a traditional discipline with an option, emphasis, specialization, or major in historic preservation.

There appears to be no unanimity as to what constitutes the core curriculum. Most programs offer a one or two-semester course that covers the scope of the field. Other frequently offered courses include architectural history and a research seminar. Beyond these courses, the core differs greatly among programs. Nearly twenty different courses offered by one or more programs are viewed as constituting "core knowledge." Because no one program offers or requires all twenty courses, clearly there is considerable divergence of opinion as to what constitutes core courses or knowledge.

Elective courses and/or those courses required for degrees in fields in which historic preservation is a specialization are added to the core courses in order to round out the usual degree program. Most degree programs also arrange internship placements where students will encounter professionals in their work. In turn, the community and the participating agency receives the benefit of student interest and assistance.

Degree programs in historic preservation are generally defined as being one of two types: (1) those which offer a degree in historic preservation, commonly referred to as a "generalist program;" (2) those which offer a degree in a more traditional discipline, i.e., history, architecture, planning, with a major, specialization, emphasis, or certificate in historic preservation, often called a "specialist program."

The proposed program from Southeast Missouri State University falls into the first of those categories since it offers a degree in historic preservation. One purpose of the proposed program is to teach students basic skills and competencies in the field of historic site administration, museum work, archival work and restoration. Also, the program will train students in research techniques and will seek to give students a broad perspective of American culture. The proposed curriculum is designed to provide students with a generalist background and a choice of four options; archives, museum, historic site, and historic restoration. The course of study contains three components: (1) the general education requirement, (2) the major (the historic preservation core, basic courses from six disciplines, courses in the option desired and field experiences in the option), and (3) the electives.

All students enrolled in the program would complete the 45-hour general education requirement for the baccalaureate degree from Southeast Missouri State University. The major portion of the proposed program consists of a historic preservation core, basic courses from six disciplines, courses in the option desired, and field experiences in the option. The historic preservation core consists of three courses (nine hours) in which students will study current efforts in maintaining past cultures, explore the values of local history, and pursue topics of particular interest.

Although the core courses acquaint students with a broad spectrum of academic specialities, the true interdisciplinarity of this program emerges in the basic courses drawn from six disciplines. Four courses (12 hours) from humanities, introduce the student to studies of American art, communication, and drawing. Students are also required eight to nine hours in history which are selected from nine courses in state, national, and folklore history. The three-hour social science requirement is selected from one of two courses which explores either the North American Indian or archaeology in North America. Students will choose between a one-hour course studying the national park system or a two-hour course in historical geography to satisfy the science requirement. The fifth course (three hours) is in housing and interior design. A three-hour course in business law completes this portion of the program.

In addition to the historic preservation core and the interdisciplinary support area, four distinct options--archives, museum, historic site, and historic restoration--are open to students. The options combine theoretical study (two courses-six hours) and field experiences (six-nine hours). Each student will devote at least two summers to the field experiences. These field experiences or internships will be organized so as to require students to draw upon their coursework and to provide a variety of experiences not available in the classroom.

Control of the internships will rest with the director of the historic preservation program. The director will recruit only those internship supervisors who are qualified, both through academic training and professional experience, to supervise the work of the interns. The university has already received offers for internships from such agencies as the Missouri Department of Natural Resources, the Tennessee Valley Authority, and the National Park Service.

A total of 24-25 hours of electives complete the 124 hours necessary for graduation.

Although the proposed program itself is new, it is composed in large part of courses which are already part of the curriculum. However, sixteen new courses will be necessary to implement the program. These include the three courses in the historic preservation core, the internship experiences, and several of the support courses.

<u>Need for Program</u>

According to the proposal, the Bachelor of Science degree in historic preservation is designed to meet societal and manpower needs, as well as to respond to student-related requests. Officials of Southeast Missouri State University indicate the program also addresses the need for diversifying the opportunities in which a well-prepared, in-place history faculty can serve the institution and the state.

The proposed program will attempt to meet needs created by a rapidly growing national and regional interest in the preservation of the past. Programs at the federal, state, and regional level, since the middle of the 1960's, have encouraged interest and growth in historic preservation. Within the state of Missouri and particularly in the southeastern region of the state, numerous communities and organizations are attempting to preserve their heritage and culture. The proposed program could provide local and regional efforts with professional assistance and needed expertise.

Another justification for the institution's investment in the degree program goes beyond the training of a small group of future professionals. University officials believe this degree will provide a unique opportunity for outreach to all ages: students, adults, and the elderly. The degree program should play an active role in the education of the adjunct community. Short courses with open enrollment; newsletters which address broad, regional interests; and the sponsorship of professional and volunteer preservationists in meetings, conferences, lectures, and forums; are but a few examples of the outreach potential presented in the program proposal.

Historic preservation, as a relatively new field, is not included in traditional occupational guides. The identification of manpower needs in this field must, as a result, be based upon individual contact with persons currently active in historic preservation work. Dr. James Huhta, Director of Middle Tennessee State University's historic preservation program, indicated to the MDHE staff that his institution has successfully placed every individual who has completed the course study. Dr. Antoinette Lee, Academic Programs Coordinator at the National Trust for Historic Preservation, informed the MDHE staff that while the demand for historic preservationists have not yet outstripped the supply, there is a steady, if not slightly growing, market for graduates of such programs. In addition, the Missouri Department of Natural Resources employed six historic site administrators in 1978-79. All six positions were filled from outside the state. Officials in the DNR indicated that no applicants for any of the positions were from Missouri.

In the proposal, Southeast Missouri State University documents both student interest in historic preservation as a profession and a demand for courses to prepare students in this field. A survey conducted by the institution in 1979 revealed that of 892 freshmen and sophomores polled, 586 indicated some degree of interest in this program. Seventy-six of those sampled indicated they were very interested in the field. In addition, the university faculty believe the historic preservation degree program will offer history-oriented students a professional opportunity other than teaching. Since the field of history is now particularly overcrowded, students who would like to major in the field are often discouraged from doing so. The proposed course of study would provide an opportunity to channel student interest in history in another career tract and effect a wise deployment of history faculty teaching strength. Furthermore, since the program will be the first in this state and region, the degree should, according to the proposal, attract students not only from the local campus, but from a broad geographic area. Based on the elements of student demand, an enrollment of eight majors is projected for the first year of the program. Projected enrollment in subsequent years is, in the opinion of the MDHE staff, optimistic. However, officials at Southeast Missouri State University indicated to the MDHE staff that they do not expect, nor will attempt to attract, large numbers of students to the program. Rather the intent of the university will be to attract interested and dedicated students, and to clearly inform them of the fluctuating job market. The MDHE staff believes this pragmatic philosophy will ensure that the program will not be producing large numbers of students who will be unable to find employment.

<u>Duplication of Program</u>

Prior to the late 1950's, preservation education consisted primarily of on-the-job training. Architects, historians, and others who entered the preservation field had little, if any, formal education in historic survey, planning, building technology, legislation, and other tools available to preservationists. In 1959, the Architecture School at the University of Virginia initiated a course in historic preservation. In 1962, Cornell University initiated courses in preservation planning. Two years later, Columbia University, through its School of Architecture, established a graduate seminar on the restoration of American buildings. By 1966-67, Columbia's seminar had formed the basis for a full-fledged graduate curriculum in historic preservation. In the fall of 1976, the National Trust for Historic Preservation conducted a survey of higher education opportunities in historic preservation which revealed that more than 20 institutions offered degree programs, and 150 other institutions of higher learning offered individual courses in historic preservation.

A preponderance of the degree offerings are at the graduate level. The first of only three undergraduate programs in the nation was established at Middle Tennessee State University in 1973 and enrolls approximately 40 majors. The other undergraduate programs are located at Roger Williams College, Bristol, Rhode Island (75 students) and at Goucher College, Baltimore, Maryland (10 students). There are at present, no historic preservation degree programs at the graduate or undergraduate level in the state of Missouri. Although a review by the MDHE staff of Missouri college and university catalogues identified a number of interdisciplinary programs and a smattering of courses that could relate to historic preservation, no institution offers the core courses considered sufficient to sustain a major in historic preservation.

Furthermore, geographic location is an important factor in augmenting the academic requirements of this type of degree program. The availability within a degree program of internships and work experiences with a variety of agencies and organizations enhances the appeal of the program's graduates. In addition, such activities give students experiences that should bolster their confidence and that of their future employers in their capabilities.

This element of locational opportunity, termed "program enrichment" by the National Trust for Historic Preservation, is central to the proposed course of study. The emergence of numerous museums in the communities of southeast Missouri which could provide employment opportunity for individuals with professional preparation in historic preservation; the development by the Southeast Missouri State University faculty of a close-working relationship with national and state agencies that are committed to historic preservation and which have expressed an interest in students as interns; as well as a diversity of preservation services to the immediate region; are all cited in the proposal as examples of why the university feels it is uniquely located to warrant the establishment of this program on its campus.

<u>Faculty Resources</u>

Because of its interdisciplinary character, the proposed program would draw from numerous academic departments for support. The university will, according to the proposal, meet projected program needs through reassignment of existing faculty. Therefore, no new or additional

-6-

faculty members will be recruited to implement the proposed program during the first four years. According to the proposal, should additional faculty be required (as the result of unforeseen circumstances) after the program is implemented, such additions will be as replacements in programs where enrollment requires less instructional support.

The MDHE staff believes the academic preparation and professional experiences of the faculty to be exceptional. Twelve of the 21 faculty members connected with this program hold the doctorate, while two have earned the MFA, the highest degree in their field--art. The remaining faculty hold the master's degree. In addition, the vita of the faculty document a diversity of professional, as well as personal activities in the preservation field. These activities include "digs" in historic sites, the restoration of historical homes in the local area, directing research projects on historical topics, collecting oral histories, and participating in historic site preservation projects. According to the proposal, the university plans for selected faculty members, e.g., from the Department of History, Library Science, to be given further training to ready them for preservation teaching. This training would not require an additional outlay of funds; it could be supported through faculty development moneys.

Therefore, the MDHE staff believe the academic preparation, the professional and practical experience, and the potential for enrichment opportunities describe a faculty capable of providing students with a quality education.

Library

According to the proposal, the program will be adequately supported by library materials, both in relation to existing courses and proposed new courses. Students enrolled in the program will have access to approximately 5,500 volumes in the area of history of the south-central states and history of the north-central states, and to 89 periodicals in the area of general and U.S. history. Library holdings from other departments which will support the program are considered adequate to meet additional needs.

The proposal cites the Kent Library Regional Collection as a special strength for the program. This collection, composed primarily of historical records and materials, reflects the economic, geographic, social, political, and educational contributions of the region. As such, it will serve as a valuable resource for potential students as well as for faculty. The library budget, according to the proposal, has provided for the acquisition of the materials essential to this program.

Physical Facilities and Equipment

The proposal indicates that classroom, library, and office space for the program are available and adequate, and no new instructional space will be needed. Field experience facilities will be provided by the organizations accepting interns at no cost to the university. Office space for the proposed program will be shared with the Regional History Center at the university.

Departments which offer courses dealing with historic preservation currently possess the equipment necessary to initiate the program. Therefore, no new major items of equipment will be required to implement the program.

Administration

Initial responsibility for administration of the proposed program will be in the Department of History and the College of Social Sciences. The program director will report to the Chairperson of the Department of History, who, in turn, will report to the Dean of the College of Social Sciences. To provide for the interdepartmental character of the program, there will be an advisory committee composed of representatives of the various departments contributing major segments to the program. To further facilitate coordination, the Dean of the College of Social Sciences will regularly report developments to the academic vice president and to the deans of the other colleges having offerings in the program.

Evaluation will rest with the program director and the historic preservation advisory committee. The director shall report these evaluation results annually to the Chairperson of the Department of History, who in turn will report to the Dean of the College of Social Sciences and the Vice President for Academic Services.

Five professionals served as external consultants to the university in the development of this degree.

Southeast Missouri State University, upon establishment of the Bachelor of Science in historic preservation, intends to become an early participant in the National Council for Preservation Education. It is anticipated that this council will emerge as the accrediting agency for educational programs in the field.

Financing

The program will be financed through a portion of student fees and state general revenue appropriations. Primary costs to the program are reflected in faculty salaries. However, faculty salaries do not represent additions of new faculty, but rather the proposed program's share of present faculty salaries in instruction and supervision. This particular program does not require any significant outlay of capital funds. The total cost should not impact significantly on the institution's budget.

AL:fht
2-28-80

[The process took another year before the program's debut. "This fall [1981] SEMO will become one of the first universities in the United States to offer an undergraduate major in historic preservation. The new degree includes a required internship in the student's area of emphasis. Students may choose from four areas: museum management, archival management, historic site administration, and historic restoration. Interns will be assigned to the University Museum, the Regional History Center, historical associations, and state and federal agencies."[13]] The restoration component later morphed into building documentation.

[Mattingly continued,] In some ways, the development and approval of the degree was the easiest part. . . . That was the start of the battle.

Initially, the Director of the Center for Regional History was also the person in charge of the Historic Preservation Program, overseeing course offerings, recruiting students, attending committee meetings, and dealing with a multitude of issues. [That was Dr. Mattingly.] Gradually, the administration of the Program became so great that the Department [of History] separated those functions, to permit more supervision of each of the two areas.

Within the Department, the Historic Preservation Program suffered from a constant lack of funds. The University also was slow to support, despite the fact that the Program attracted students from across the country.

...One of my really tough fights was to find a way to let people know that we existed. Because you don't know we exist there isn't much of a chance of getting people in to participate in the program. So, one of the big fights was to get a mailing list of high school seniors across the United States that were interested in history in some area they thought they were interested in. Maybe outdoors work, forestry, maybe the park service.

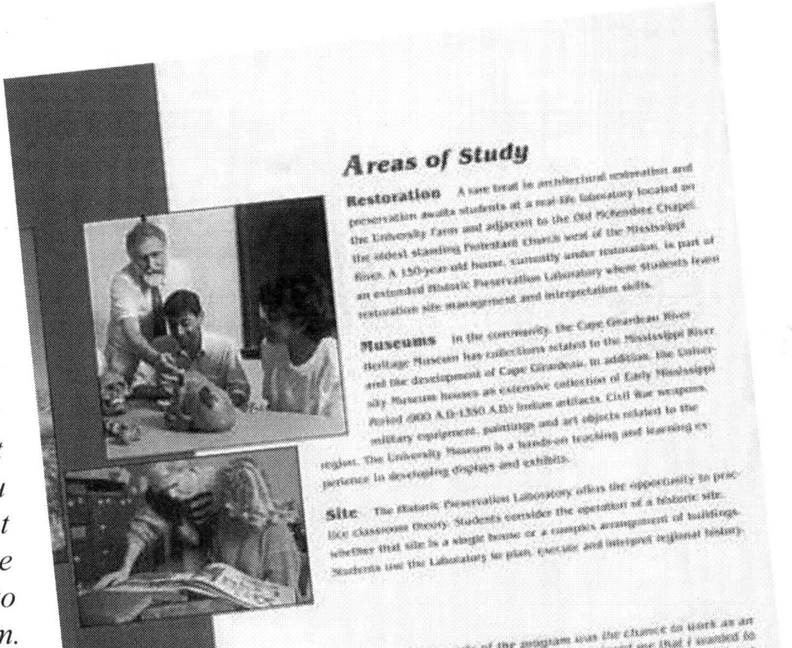

One panel from the first informational brochure for the SEMO HP Program, directed to prospective students from across the country.

Then I would send them a letter. And we began to get inquiries back from people. Well, I want you to come to campus, so I can talk with you. Bring your mom, bring your dad, bring one of your parents with you because I want to meet them and talk with them. That's how we began to build the program. People began to find out about us.

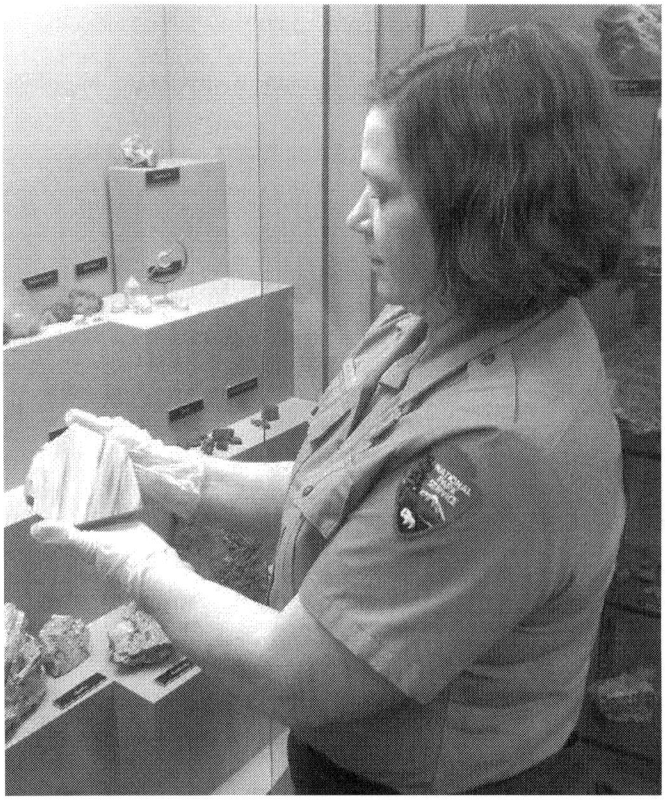

1986 graduate Jackie Holt on the job with the National Park Service in 2016

And as a part of this degree program, one of the things I insisted on having a was a semester internship. That was required of everybody regardless of what part of historic preservation you were going to be in. They had to give up a year here on campus to go someplace and practice the area that they felt like they were interested in. Museum work, site administration, historic downtown, whatever that was what they were going to have to do.

Now, it's pretty common to have internships, but 30 years ago this was a new kind of thing and it gave me an opportunity, at my expense, to travel wherever these interns were and personally visit them as well as the people who were running it.

So, I got to see the country. But it was also important because it let you know that at some point in your internship, that during your summer you would get a visit by someone, usually me, in the program. And the site where you were working knew that I was going to come and visit. You wouldn't know, but they would know. I didn't want the intern to know. So, it was a way to visit and know how the intern was doing and how things were going.

It was also publicity because the place where the student was would see me; see the kind of product we were working on. If they did good work, they might even get a job there one day. And they might be willing to take another intern or two as timed progressed. It worked out very well. And I got to see many places and we got students all over the United States. Just great.

^ HP student Randy Greeves cataloging artifacts, 2020

As one of the individuals involved from the beginning it has been very rewarding for me to see the Program continue to grow and attract high quality academic majors. I believe the Program still follows its original goal of offering an academic alternative to those who like history/cultural heritage and want to teach in a setting other than the classroom. [end of Mattingly's 2006 recap and 2010 oral history interview.]

^ Dr. Art Mattingly (left) and
Dr. Steve Hoffman (right),
HPA Banquet, 2006

> 1982 graduate Denise Morrison
at the Kansas City Museum in 2021

Putting Pieces Of The Past Together For The Future

by Pam Dinsmore

Most people would never dream about dismantling a house only to reconstruct it somewhere else. But a handful of students on campus participated in just that — the reconstruction of a log house on the university farm.

These students belonged to the Historic Preservation Association. This group was formed by those students who were historic preservation majors.

A search was conducted in the spring of 1982 for a house for the university's historical farm. The house was found in Sikeston and donated by Mr. and Mrs. James Moore to the university. As the house was being taken apart, each log had to be numbered and tagged so it could be reconstructed correctly. "This was the first stage of building a living historical farm," said Lisa Couch, senior historic preservation major. "We are also planning to acquire a barn, corncrib, and other buildings and position them in certain areas around the farm, so it will look like the farms did in the mid 19th century."

Students involved in HPA or wanting to obtain a historic preservation major in site administration, museum, or archives were required to get practical experience by working on the log house.

Just what is it that HPA does? Members preserve, naturally. Confusion arises when discussing preserving vs. restoring. Restoring involves reestablishing a site or structure to its natural composition whereas preserving is taking steps to keep it the way it is.

During the fall, two members, Jesse Francis and Debbie Reid, went to Marietta, Ohio, for a Pioneer of America meeting. "We were the only undergraduates there, and it made us feel good to be able to present a paper on our work," said Francis.

Problems and progress of the log house was the main topic of the paper they presented. Reid focused on planning the move of the log house and how they transported it from Sikeston to Cape Girardeau. The group hoped to make people aware of its goals and what members were doing to accomplish them.

Exhibits were always on display in the University Museum in Memorial Hall. They showed past cultures by the different kinds of dress they wore and tools and utensils they used.

Another reserved place for exhibiting was the Missourian Room on the third floor in Kent Library. This room contained the Regional History Collection of old books, papers, manuscripts, and even tax records. "We try to put the archives we've collected into some kind of order," said Lisa Couch.

The HPA participated in Homecoming by entering a float displaying an old steam-engine tractor. Guest speakers also highlighted the year for the group. They invited speakers that had a background in historic preservation or architecture.

The members had one main goal in mind. They wanted "to encourage the recognition and preservation of artifacts, sites, structures, and cultures within the community and surrounding area," Couch said.

Historic Preservation Association officers were: Debbie Reid, president; Mary Malone, vice president; Karen Luker, treasurer; Lisa Couch, secretary; Melanie Gimpel, historian; and Dr. Arthur Mattingly, advisor.

ON SEMO'S UNIVERSITY FARM, the first log was laid to begin the reconstruction of the B. F. Hunter log house. Karen Luker and Scott Taylor, along with other historic preservation students and faculty, place the log on its final foundation.

1983 *Sagamore* yearbook

140 Historic Preservation Association

Bob Hulsey

A COLLECTION OF authentic World War II attire donated by L. B. Twaddell of Benton was displayed in the University Museum during October as part of HPA's emphasis on our past. Mary Malone puts the finishing touches on this mannequin dressed in a Women's Auxiliary Corps uniform.

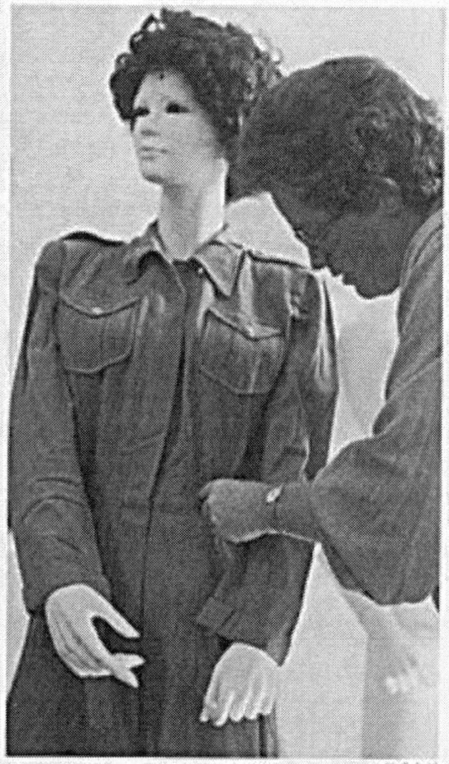

Mark Smith

Bob Hulsey

HISTORIC PRESERVATION ASSOCIATION: **Front Row,** Dr. Arthur Mattingly, Lisa Couch, Robin Richardson, Karen Luker, Claudia Finley, Debra Reid and LaCretia Drum. **Back Row,** Bill Lee, Frank Woodson, Jesse Francis, Robert Ulrich, Melanie Gimpel, Ralph Seyer and Mary Malone.

Organizations **141**

Birds Eye View State Normal School
Cape Girardeau, Mo.

Where the past,
MEETS THE PRESENT.
SOUTHEAST MISSOURI
STATE UNIVERSITY · 1873

2

A Statement of Beliefs

SEMO HP Program Coordinator Dr. Bonnie Stepenoff described the Historic Preservation Program for prospective students on the Department of History's website in 2001-2002. No longer accessible on the Department's website, the following was mined on the "Way Back Machine" at archive.org:

The purpose of the program is to develop skills in research, interpretation, and curation of landscapes, buildings, objects, and documents of historical significance. The Department of History offers this program to prepare students for employment in historic sites, museums, archival repositories, or preservation agencies. Graduates of the program may also pursue advanced study in public history and historic preservation.

By the time you complete the program, the Southeast Missouri State University Department of History believes that you should:

* Develop a knowledge base in history.

* Develop skills in historical research, including the use of appropriate computer databases.

* Develop skills in locating and using primary and secondary local history sources, which will enable you to prepare historic surveys, National Register nominations, museum exhibit text, and historic site interpretation.

* Develop skills necessary for the identification and documentation of historic buildings and objects.

* Develop skills in oral and written communication.

* Develop basic skills in measuring and drawing buildings and objects, including skills in the use of Computer Assisted Drafting (CAD) programs.

* Develop a basic knowledge of legal and economic issues pertaining to historic preservation.

* Develop specific competence in one of the following areas: archives and special collections management, museum management, or historic site administration.

* You will practice classroom theories by completing an internship at a local, regional, or national site, providing hands-on experience in varying facets of museums, archives, and site administration.

* Develop knowledge and skills in a field that complements your work in Historic Preservation.

* Develop basic skills in the use of computers for word processing, database management, and drafting.

Students in Historic Preservation attending an interpretation workshop at Mount Carroll Illinois, March 1985.

Field trips and bus tours have always provided experiential learning for HP students, *Pastscript* newsletter, 1984-1985

To assess your progress in completing the program, the Department of History will compile an assessment file. One important part of that file will be your portfolio. Creating a portfolio will help you to evaluate your own progress and to demonstrate your accomplishments to others.

You will be required to complete History I, History II and nine additional credit hours in history as specified in the curriculum. One part of your American History I (US105) course will be the administration of a historical awareness test. This instrument will act as a benchmark for you and help us assess where you are in meeting the objectives. During your senior year you will be given a similar test so that we can chart your progress more effectively.

Instruction and practice in research methods will be incorporated into several courses, including HP100 (Introduction to Historic Preservation), HP200 (Techniques of Local History), HP405 (Archives and Special Collections Management), HP410 (Museum

Studies), HP420 (Historic Site Administration), and HP588 (Legal and Economic Principles of Preservation). Project reports from these classes will be part of your portfolio.

In order to develop specific skills relating to historic preservation, you will be required to complete HP200 (Techniques of Local History). In this course, you will complete research projects that will be included in your portfolio.

You will learn about buildings in HP585 (American Architecture). In order to become familiar with historic properties and objects, you will participate in field exercises, such as historical/architectural surveys incorporated into their course work. Field work will be an essential component of HP417 (Problems in Applied Museum Studies), HP425 (Problems in Site Administration), and HP588 (Legal and Economic Principles of Historic Preservation). Reports on field work will become part of your portfolio.

You will demonstrate communications skills by writing reports, giving oral presentations, participating in group projects that require writing and speaking skills, and finally, submitting examples of written work in a portfolio.

^ Hunter Log Cabin, ca. 1994

< David W. Jackson's May 1993 perspective from Dr. Chamberlain's AR207 Practical Drawing

You will demonstrate graphics skills by successfully completing AR 207 (Practical Drawing) or VC100 (Introduction to Drafting) and/or VC126 (Computer-Assisted Drafting) and by submitting examples of drafting in a portfolio.

You will be required to complete HP588 (Legal and Economic Principles of Historic Preservation).

You will demonstrate competence in an area of preservation by completing one 400-level "studies" course and one 400-level "problems" course in one of these areas.

You will further develop your skills by successfully completing an internship.

You will discuss your internship in a public presentation.

You will develop additional skills by completing a minor in an area such as History, Archaeology, Architectural Design, or Promotion Management.

You will be asked to demonstrate computer competence. You can do this by completing AD101 (Introduction to Microcomputer Applications) or by presenting convincing evidence of computer skills to an appropriate faculty member and by including work completed with the aid of computers in a portfolio.

Members of the faculty will evaluate a portfolio, which will include:

(a) a resume.
(b) a current course of study form indicating courses completed.
(c) copies of all project reports completed during field exercises.
(d) a summary of contributions to preservation projects
(e) other evidence of competence in preservation.

You will discuss your portfolio in an exit interview.

< 1989 graduate Susan Haake protecting Abraham and Mary Lincoln Home National Historic Site in Springfield, Illinois, 2016

>1993 graduate Tony Ford during his internship at the Arthur Bowring Sandhills Ranch State Park near Merriman, Nebraska, 1991

3

Center for Regional History and Cultural Heritage

"This eventful year on campus [1980-1981] also included the opening of the Southeast Missouri Regional History Center [established in 1978[14]]. Currently [1981] housed in Kent Library, the Center is under the direction of Dr. Arthur Mattingly of the Department of History. It includes sets of photographs, newspaper files, and historic university publications dating back to the late nineteenth century. It also holds the legislative papers of Vernon Bruckerhoff, member of the Missouri House of Representatives; and the papers and correspondence of [Julien N.] Friant, undersecretary of agriculture in the administration of Franklin Roosevelt. All records are kept in acid-free folders in temperature-controlled rooms.

"A special feature of the Center is a collection of oral history interviews. They include those conducted by Dr. Mattingly [*Hermann: The Rhine City on the Missouri River*]; by Mr. Robert White among Mississippi riverboat captains and crew members; by Herb Poston concerning the World War II prison camp at Weingarten, Missouri [Weingarten Prisoner of War Camp Collection]; and Dr. Paul Travis with physicians and local residents concerning folk medicine in Southeast Missouri [i.e., Colonel John Buxton]."[15]

These materials became part of SEMO's Special Collections and Archives, finding aids for which are online, together with more than 500 other, distinct, regional history collections.

The Center for Regional History and Cultural Heritage serves the University and the region through its work in documenting, preserving, and presenting various aspects of local and regional history; architectural trends; archaeology; ethnic and folk traditions.

Members of the Center staff provide assistance to museums, historical societies, schools, and a number of city, county, state, and national agencies.

The Center offers a wide variety of consulting services, including museum registration and cataloging, cultural resource management, historic architectural surveys, as well as program planning in regional and local history.

The Center is also involved in public outreach activities which serve to promote the value of historic preservation and regional cultural expression.

The center is administered within the College of Liberal Arts and serves as a support unit for student and faculty work in history, historic preservation, anthropology, archaeology, and folklore, as well as other fields.

Coordination of various programs which serve targeted age groups as well as the general public. These include:

Field Projects.

In this broad category the Center refers inquires and requests to appropriate and/or interested faculty and departments. Recent projects include the development of a battlefield protection plan for the Fort Davidson Civil War site at Pilot Knob. Additional projects have been developed in Ste. Genevieve; New Madrid; Chaffee; Marble Hill; Bonne Terre; and Jackson. An extensive historical survey of the Whitewater Creek in southeast Missouri was recently accomplished for the Missouri Board on Geographic Names.

Public Service.

The Center for Regional History serves a significant public service function. Local, regional, and state-wide groups call to request speakers, programs, and information. Dr. Frank Nickell made countless speeches on topics of interest to various organizations. The Center for Regional History plays a role in "taking the university into the region."

Publication.

Another way the Center reaches those it serves is through articles in various publications. By 2010, more than 25 local and regional histories were published, and others in process:

Advance, Missouri: A Look at the First One Hundred Years by Miles Moroni
Around the Town of Cape Girardeau in Eighty Years by Jo Ann Bock (2013)
Back Woods Jazz in the Twenties by Peg Meyer (1988-1989)
Battle of Pilot Knob, Mo. by Bryce Sudero (1984-1985)
Big Muddy: A Journal of the Mississippi River Valley, Volume 1, Number 1
Bondage in Egypt: Slavery in Illinois by Darrel Dexter (2010)
Canalou: People Culture Bootheel Town by Dan Whittle
Cape Girardeau Democrat Index, 1891-1895
Cape Girardeau: Then and Now by Tom Neumeyer (2010)
Child of an Eagle, The by William Robert (Bob) White.
Reprint dealing with Mississippi steamboating (1986-1987)
Cultural Resources Survey: Apple Creek, Mo. (1984-1985)
Cultural Resources Survey: Cape Girardeau Port Authority (1984-1985)
French and Germans in the Mississippi Valley: Landscape and Cultural Traditions by Michael Roark
From Farm to Front: An Innocent Goes to War, 1942-1946 by Lawrence E. Breeze (2009)
Good and Ordinary Life, A: A Memoir of the Rural Midwest by Joe Dunn (2013)

History of the Saxon Lutheran Immigration to East Perry, Mo.
Reprint of Gotthold Heinrich Loeber's 1839 German language original. (1984-1985)
History of Trinity Lutheran Church (1988-1989)
House on Riddle Hill, The by Glenn Tompkins (1996-1997)
Momma Don't Ya Want Me to Learn Nothin?
Reflections of a Young Man's Life by Eugene M. Munger, Jr. (2009)
Mosaic of Memories, A: Of the Alumni Merit Award Recipients by Jane Cooper Stacy (2011)
Rural Schools and Communities in Cape Girardeau County by Lacy Christabel and Bob White (1986-1987)
Tears of the Trail by Marie W. Exler
Thunder in Arcadia Valley: Price's Defeat, September 27, 1864 by Bryce A. Suderow
Twelve Corners: The Refuge of My Youth by Ruby Allee Wright (2020)
Vietnam: Angel of Death by Harry Spiller (2002)
Voices of Bunker and Extended Communities, The: Centennial Edition, 1907-2007 by E. Dean Burns
Way it Was in Southeast Missouri, The by Luella Duncan (2013)

Other Center for Regional History Accomplishments

Academic Hall print by Jake Wells

Mississippi River Tales Mural, Cape Girardeau, Missouri

Inventory and catalog artifacts at the Towosahgy State Park (1986-1987) (completed 1988-1989)

Oral History Projects: Wartime experiences of area veterans of World War II (1984-1985)

Oral History Project: Demise of the cotton industry in Southeast Missouri (1984-1985)

Study of French and German settlement in the Mississippi River Valley (1986-1987)

Sponsored a conference on downtown revitalization, sent via. satellite from Washington, and Center personnel serving as moderators in Cape Girardeau, while viewers were able to participate by telephone in the discussions. (1984-1985)

Sponsored the Annual Mississippi River Valley Scenic Drive beginning in 1991-1992, the fourth weekend each April. The 19th annual Mississippi Valley Scenic Drive was in 2009.[16] The April 2021 event was canceled; but, they are planning for their 30th Anniversary scenic drive in April 2022. (facebook.com/MississippiRiverScenicDrive/)

Sponsored the annual Veryl L. Riddle Distinguished History Lecture. The 8th Annual was on 12 Nov 2008.[17]

Launched and sponsor of the blog *SEMO HISTORIAN: Where History Meets the Present.* "Our blog is aimed to show what our History and Historic Preservation students, professors and alumni are contributing to the field. Here at Southeast Missouri State University our small class sizes and award-winning faculty mean that you will receive the personal attention and warm community support that become the keys to future success in the field of History and Historic Preservation." (semohistory.org) (launched 7 Oct 2018)

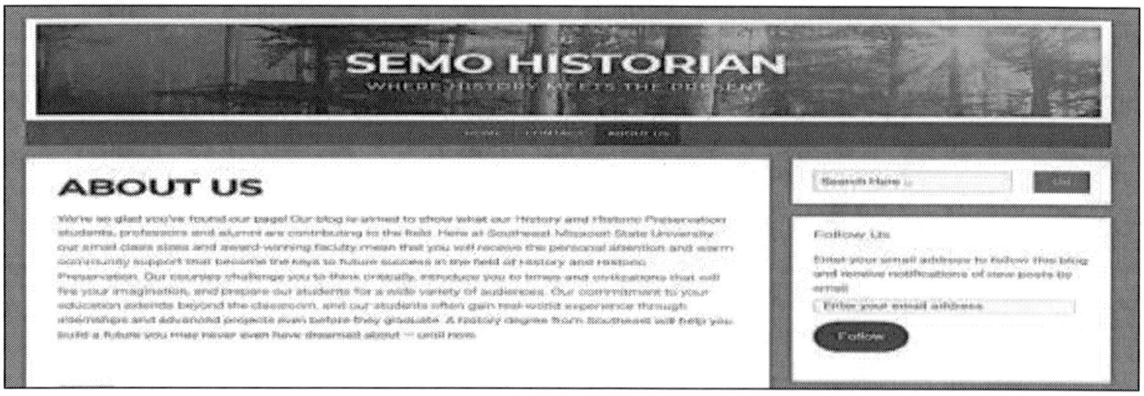

Center as a Clubhouse to HP Students

By 1991, the Center was located at 520 North Pacific, a former private residence built in 1936 by Renard and Lenore (Nora) Walther.[18]

HP students could come and go, mix and mingle, or just find a peaceful respite between classes at the Center. "*It allowed for a real sense of community among HP students*," said David W. Jackson (1993 BSHP; his AR207 Practical Drawing perspective is below). "*There were all kinds of friendly kids of all stripes happily spending quality time together at all hours of the day.*"

Later he added, "*It was what woke youth in 2021 might call 'boho'.... And then there was Sue Corvick who greeted you and brightened your day with a great smile every time you entered.*"

Robert "Bob" White succeeded Mattingly as Director of the Center during the 1983-1984 academic year.[19]

The Center received $4,000 from the Cape Girardeau County Court to microfilm the records of the County Clerk, 1828-1897. The Center also acquired a photo-negative collection from Bollinger County, Missouri, of ca. 1900 images of rural scenes, timber operations and drainage projects. Additionally, the Center acquired the Roberts Military Collections of World War II correspondence and campaign maps.[20]

Dr. Frank Nickell became the Director for the Center in 1990-1991.[21] "For more than twenty years, Dr. Nickell served as the Director and became synonymous with the history of southeast Missouri."

Dr. Frank Nickell and Cleo, one of two resident cats at the Center for Regional History

1998 BSHP graduate Anne (Kern) Bauer shared the above photo with this recollection:
"I took B & W Photography in 1997. For my assignment on lighting, I took a photograph of Dr. Nickell at his desk with Cleo lying on it. It was so dimly lit; I didn't know if it would come out. It was one of my favorite pictures I've taken. The negative is long gone, but I gave a print to Dr. Nickell. At the beginning of my freshman year, I was a little afraid of going in there, but once I did, I found my home away from home. I don't know of any other major on campus with that kind of space."

Spring 1997 documents show the study of "re-use of University owned property at 520 N. Pacific."[22] In the later 1990s, the Center for Regional History was relocated to the third floor of Carnahan Hall (Room 301).

In November 1999, with help from the Catholic diocese, Fr. John Friedel arranged for purchase of 520 N. Pacific.[23] They razed both their original house and the former Center for Reginal History to build the new Marquette Student Center, which the Catholic Campus Ministry relayed on its website as of September 2021: "We are temporarily located at 902 College Hill for the construction of our new building. We were beyond blessed to have raised $5.2 million for this project that is now underway. We anticipate the construction to be completed in early to mid-2021."[24]

The HP student body lost a vital element to their daily experience at SEMO. Their opportunity to casually gather on campus in a freestanding, autonomous location equal to other social groups like fraternities' and sororities' houses was lost. That neutral, shared space where students could meet and balance their academic and social lives was no more.

In recent years, a couple of other older homes along Pacific have been spotted as a potential future "HP house" that is within walking distance of campus (i.e., campus adjacent). *Such a dream will take some political maneuvering…and money…to bring into reality.* One location, 831 North St (southeast corner of Pacific and North), has been on the Cape Girardeau Historic Preservation Commission's endangered buildings list.[25]

< Another house sold recently at 857 Alta Vista Dr could have been a perfect "HP House" but was, perhaps, too great a vision.

Any campus-adjacent house could be restored and maintained *by HP students* honing their professional skills. Hands-on field experience close to school…plus practicum credit? And, a restored campus-adjacent house and HP student clubhouse?
Win. Win. Win!

Is there another home nearby that might one day serve as a communal gathering "HP house?" Keep your eyes open. Prepare to mount a funding and feasibility strategy!

Though the home at 520 N Pacific is literally history, living memories persist. These quotes were gathered from a SEMO HP Alumni Facebook survey in June 2021 that reached 213 people, enjoyed 97 engagements, and garnered comments like those aforementioned, plus:

> *"Bart and Cleo [resident cats], Dr. Nickell, Soup on Wednesday, coffee and cookies, Sue Corvick, HPA meetings. The Center for Regional History was the place to go between classes to talk, plan, think, and respite."*
> –Sandra (Shupp) Campbell (BSHP, 1996)

> *"I believe the Center for Regional History on Pacific came about while I was in the program. It was a grand American Foursquare, and a place to spend time with kindred souls, especially on soup days!"*
> –Creager Smith (BSHP, 1989)

> *"Working on a backdrop for an exhibit in New Madrid. We were told we could use the basement. We went down there and discovered human remains laid out on a table (turns out they were being repatriated). We painted our backdrops in the basement. We came back next day to find Bart had walked across them."* –Stacy Reaves, Ph.D. (BSHP, 1995)

> *"A few of us made it our mission to stump Dr. Nickell on obscure trivia, but it was damn near impossible! I happen to be taking a Shakespeare class reading comedies and histories and got him on a footnote in Henry IV. I remember Jon Colburn* [BSHP, 1995] *made a wreath out of punch cards and painted it silver to match the aluminum tree for the Christmas party one year. I think Mary Ann Kellerman has that wreath."*
> –Danielle Lewon (BSHP, 1996)
> [to which Dr. Kellerman replied: *"YES!! I treasure that wreath. We use it as Heritage Hall at Christmas. Thanks Jon!!!"*]

By 2012, the Center's name was changed to the Lynn S. Bollinger Center for Regional History.

The Center's Facebook page was inaugurated on June 20, 2012 (facebook.com/CenterForRegionalHistory/).

In 2013, following the retirement of Dr. Nickell, **Dr. Adam Criblez**, Professor of History, became Director of the Center and served through August 8, 2021.

^ Dr. Adam Criblez

> Dr. Steven Hoffman

Dr. Steven Hoffman was named director of the Bollinger Center for Regional History and Culture and assumed his duties on August 9, 2021.

The Center has come full circle once again. Whereas in the beginning when Dr. Mattingly was both the HP Program Coordinator and the Director of the Center for Regional History, so now is Dr. Hoffman. Congratulations and best wishes, Dr. Hoffman!

"I look forward most to the opportunities to engage students and faculty in the University's Historic Preservation and Public History programs with enhanced experiential learning opportunities and outreach in the community," Hoffman said, adding, *"My main goal as director will be cooperating with area institutions to strengthen the teaching and scholarly opportunities for preservation students and faculty, providing strong learning experiences for our students, and serving the needs of the region."*[26]

When interviewed by the Southeast Missourian the previous year on the eve of the COVID-19 pandemic that disrupted everyday life, Dr. Hoffman answered the question, "What is on your mind in regards to historic preservation in Cape Girardeau right now?" He answered:

> *"We have these buildings that have value, even if we don't necessarily see it today. And once they're gone, they're gone. I am a strong advocate for if we're going to build today, we should build quality, and we should build something that is going to be worth preserving and will not only be around in 50 or 75 years but is something that people will want to*

preserve. I was very pleased with the direction the redevelopment of the Courthouse of Common Pleas seems to be taking in that there does seem to be a commitment to quality that will be representative of the quality of our community and an investment in the future. They're not going to just look at the bottom line and build the cheapest thing, but in fact they're going to build something that we can be proud of and recognize the history that's there and what's needed to get there with some new construction. I think it should speak to the age that we are in. Let's not build something that looks old; let's build something that fits in with what we have but looks like today."[27]

As of October 2021, the Center's webpage stated:

"The Bollinger Center for Regional History at Southeast Missouri State University is involved in a variety of public outreach programs which serve to preserve, present, and promote the history of the region.

"The Bollinger Center for Regional History is located on the third floor of Carnahan Hall (Room 301) on the campus of Southeast Missouri State University in Cape Girardeau, Missouri.

"The Center fills many roles both on- and off-campus. We advise and assist students working on scholarly projects focused on the region and present topics of local interest to various classes at Southeast.

"For the broader community, we provide tours and lectures to civic and educational organizations about the history of the region. We also collaborate with local museums and historical societies to promote local and regional history."[28]

Carnahan Hall

4

HP Experiential Learning

Every student in the SEMO HP Program is expected to learn and grow through experiential opportunities, generally focused on their expressed area of interest in historic preservation (Historic Site Administration, Museum Studies, and Archival Management).[29]

This chapter explores a variety of experiences that HP students have been exposed to over the last 40 years. Some projects and classes continue to this day; others, experimental, proved to have less endurance. Files for many student projects are archived in the Special Collections and Archives Department in Kent Library.

The earliest, first, hands-on experience was the restoration of the Hunter log house.

Benjamin J. Hunter Log House

Art Mattingly described the beginning of the Hunter House project:

"Early on . . . once the degree was approved, one of the things I wanted to do was to have a project that would involve the early comers so that they would buy into the program. So that they feel a part of this initial character. And one of the things we toyed with- there was an announcement in the paper that the government would give away a log house. And the reason they were giving it away was they were building a lake in northeast Missouri that ultimately became Truman Lake, I think. All you had to do was come get it. It was yours and you could do whatever you wanted with it. Well, that was not possible, totally impossible to do because it would require us to either a) move it out to the river and float it down the Mississippi, which would have made a good program on national television or something. It was just not a practical idea for us, but we never gave up the idea to get something that would involve the early students.

One afternoon I was sitting in my office and in came one of my students in HP. He was breathless. He said, "You've got to stop whatever you are doing and come NOW and see this house that I have found."

Because these early students would get out on their own and wander around the countryside. All sorts. I said I cannot go right now, but we'll go. The short of the long is he showed me a log house. I happened to know the persons who owned the house [Mr. and Mrs. James (Handy) Moore]. I had grownup across the street from the individual.

Over a period of time, we opened negotiations with these people, and they agreed to donate the house to the university.

The University had to find a place for it. Once the University found a place for it, and I asked the students in the spring semester of my HP 100 class, I asked them if they would give up a summer and volunteer, they would get college credit, to help dismantle, tag, and move this house. And they agreed.

And that starts the beginning of the Hunter House."

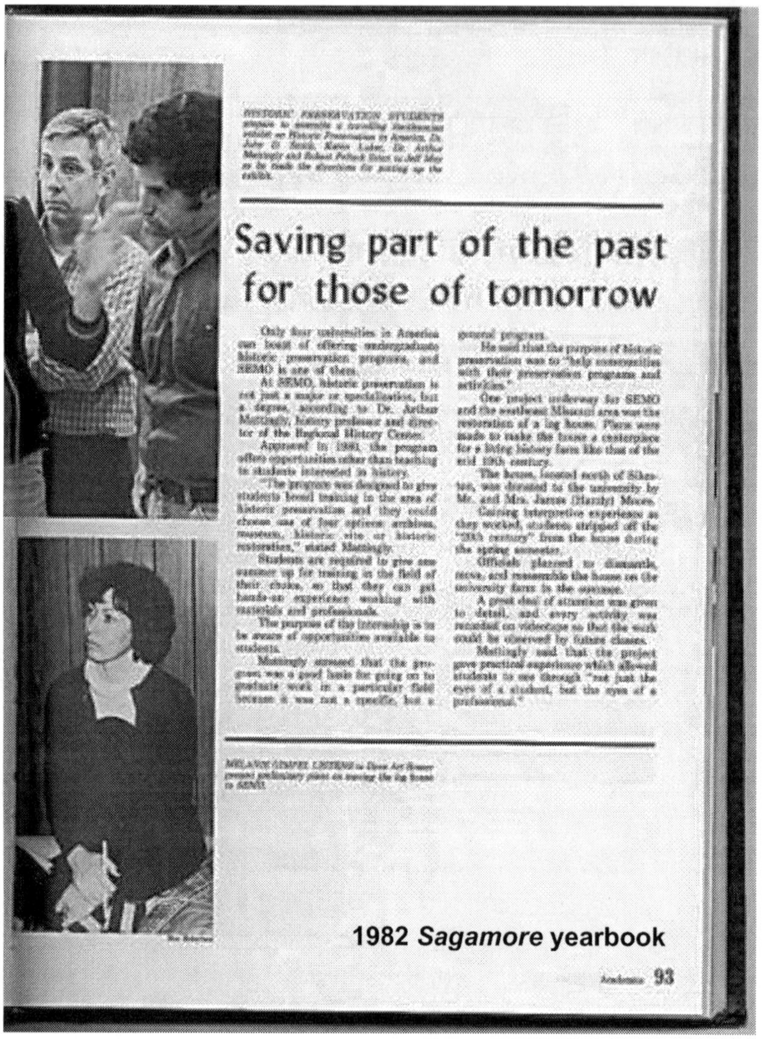

Saving part of the past for those of tomorrow

Only four universities in America can boast of offering undergraduate historic preservation programs, and SEMO is one of them.

At SEMO, historic preservation is not just a major or specialization, but a degree, according to Dr. Arthur Mattingly, history professor and director of the Regional History Center.

Approved in 1981, the program offers opportunities other than teaching to students interested in history.

"The program was designed to give students broad training in the area of historic preservation and they could choose one of four options: archives, museum, historic site or historic restoration," stated Mattingly.

Students are required to give one summer up for training in the field of their choice, so that they can get hands-on experience working with materials and professionals.

The purpose of the internship is to be aware of opportunities available to students.

Mattingly stressed that the program was a good basis for going on to graduate work in a particular field because it was not a specific, but a general program.

He said that the purpose of historic preservation was to "help communities with their preservation programs and activities."

One project underway for SEMO and the southeast Missouri area was the restoration of a log house. Plans were made to make the house a centerpiece for a living history farm like that of the mid 19th century.

The house, located north of Sikeston, was donated to the university by Mr. and Mrs. James [illegible].

Gaining interpretive experience as they worked, students stripped off the "20th century" from the house during the spring semester.

Officials planned to dismantle, move, and reassemble the house on the university farm in the summer.

A great deal of attention was given to detail, and every activity was recorded on videotape so that the work could be observed by future classes.

Mattingly said that the project gave practical experience which allowed students to see through "not just the eyes of a student, but the eyes of a professional."

1982 *Sagamore* yearbook

Academics 93

The Hunter House project continued during the 1983-1984 school year. In 1984-1985, Dr. Mattingly reported that the Hunter House project was nearing completion. Further details and interpretive additions were reported in 1986-1987. In 1992-1993, student's, under the direction of Dr. Mattingly, planned and conducted restoration work on an 1880s log smoke house which previously had been dismantled and moved to the Hunter Log House site. (Sources: *Pastscript: News of the History Department*, Number 9, 1983-1984, including a full-page of images; Number 10, 1984-1985; Number 12, 1986-1987; and, Number 18, 1992-1993.)

RECONSTRUCTION OF THE HUNTER HOUSE

A beginning is made.

Historic Preservation students receive "hands on" training.

The walls of the Hunter House nearly completed.

Detail of notched logs at the corner.

Detail of interior corner.

Chinking with clay after the logs have been set.

The house was to be the centerpiece for those interested in historic site administration, who were interested in museum work, who were interested in period costume, who were interested in teaching elementary school.

What we wanted to do was bring elementary schools- have part of their curriculum to bring them to this house and involve them in the different crafts that were used in the 1850s. The house was built by slaves before the Civil War. They had Cyprus timber and it was a magnificent house. So, that was our vision for the house, it was to be the centerpiece. We were going to move a smokehouse in, a barn, all these things. But in order to do these things, unfortunately, you have to have money.

The HP Program was housed in the history department. And people in the history department loathed to give money to the preservation program that they felt necessary for their library books and their this and their that, which was perfectly understandable.

Erecting the scaffolding.

Preparing to dismantle the smokehouse.

Dismantling the smokehouse.

So, here we are, two or three of us trying to get money for some of the things we did. We moved the house without any help whatsoever from this university. All of these things we got because we went out to the people and asked, "Would you help us?"

Getting the Cyprus tree because one of the main foundation logs was gone. We found a man who gave us a Cyprus tree. We found a man who cut trees down professionally and hauled them out. He brought in a big machine that cut the Cyprus tree down. Hauled it out for us. We brought it to where the Hunter House is today [2010]. *The log, just the log, the log was over 16 feet long. The problem was that most sawmills do not cut anything longer that eight feet or ten feet. We had to find a mill that would essentially custom cut it for us. We could then use chip and so it would look like original. That is one story.*

[Involved at the outset were] *myself, Bob White, a woman who lived in Jackson who had a degree from the university who was looking for taking classes in the master's program. Not preservation, we barely had a preservation program at that time. Her name was Lucretia Drum,*[30] *and she was so important to us because she knew a lot of people in the community. And she would go out to them and say would you do this, would you do that, would you help us? Penzol Construction, for example, came out and shot footings for us so we would have the exact positions and accurate holes. And she did things like that. And Art Rosser who was industrial technical education also helped us.*

It was wonderful. That was the best, worst mistake that they ever made.

That was great for us.... In this whole process of taking the house apart, tag it, moving it and putting it back together, we never got so much as one mashed finger....

So, the students would go some afternoons and the history department gave me a class that had a day designated, and the students would sign up- get college credit for it and we would go down and work on the house that day. So that's part of how we did it. We would spend the weekend working on the house. I made no distinction on working on the house, whether you were man, woman or child you would do the same thing.

And I shall never forget one spring say we were upstairs in the house taking off the wallcoverings and I heard a young lady scream to the high heavens. And guess what she found behind the wall? Snakes nest. Black snakes. So, we caught them and carried them out of the house and into the field so they wouldn't have to kill them because black snakes are very good. I mean they really are, you know? I know, I don't like snakes particularly either, but I want you to know we were back down there the next weekend. They were back. They were back. But my point here is that everyone in the program was treated equally and it made no difference. We all wore hard hats, we all wore gloves, we all wore boots. We all did the same things. I worked right along. I swept, I had difficulties like everyone else, but we made no distinguishes. It was wonderful. And, I couldn't have done it if they hadn't have done it. That was the key to it and again it was an opportunity for them, too. For this program and these people who were in the program. When I see them occasionally they always say, "what about my house?"

...There is blood, sweat and tears in that house. Not only from the slaves that originally cut the timber and put the house up, but the other slaves that came along in the 20th century and moved it....

Oftentimes, in the summer, we would stop at a watering hole and sit down and critique what we had done that day. What mistakes have we made? And every day before we would go we would sit around and talk about what we were going to do today. How are we going to do this, how are we going to do that? Who is designated to take the photographs today? Who designated to do the video? And that kind of thing. And we always looked after each other. So, it became almost a family and it was an experience I will cherish always.

In August 1996, it was reported that the Hunger Log House had a new look. "University crews cleaned up the grounds and installed a security fence and gate. Now it's time to get to work on the preservation maintenance of the house."[31] On November 23, "fifteen students and two faculty members accompanied architect, W. Philip Cotton, Jr., on a field trip to the house.... Teams of students sketched and photographed the property. Others took field notes and recorded the architect's comments on the condition of the log structure. Basically, the cypress log structure is sound, but suffering from water damage. There is an immediate need to cover the roof, fill window openings, chink and daub the structure. Grading will be necessary to route water away from the huge sill logs that support the structure. During the spring semester, HP students will work with Mr. Cotton to study preservation options and write a plan for the long-term maintenance of the property. After the field exercise, 23 students and faculty members from the historic preservation and interior design programs met with Mr. Cotton at the Johnson Faculty Center to reflect on the role of the architect in preservation and the ways in which architects, historians, and design professionals can work together on preservation projects...."[32]

In fall 1999, under Brian Driscoll's leadership, "students re-glazed all the windows, moved earth away from sill logs, and treated structural members with borates to prevent further deterioration due to fungus and insect infestation. Thanks to HPA president Donna Rausch there are now picnic tables on the Hunter House property, where we hope to hold future events and class sessions. Other recent improvements to the property included landscaping to improve drainage and plantings to enhance the appearance of the site. Roof repairs are desperately needed. This property is important to the Historic Preservation Program as a learning laboratory. We need everyone's help and support to ensure that it is maintained for the future."[33]

A decade later, however, the Hunter log house was in sad shape by 2010. When SEMO sold the University farm in the intervening years, the Hunter log house was donated to the Old McKendree Chapel, a national Methodist shrine dating to 1819.[34]

McKendree was soon to reach its bicentennial year when Mark Rademaker and his family moved back to Cape from California.

Mark and Tia Rademaker were both raised in Cape Girardeau where they met downtown amid the historic buildings and live music.

Mr. Rademaker said, *"After watching too much 'Barnwood Builders' on the DYI cable TV network, we began creating a barn wedding venue called Rusted Route Farms at 5739 State Highway W, in Jackson, Cape Girardeau County, Missouri. He continued, "One day in late fall 2017, I was driving up the Interstate and because the leaves had fallen from the trees by then, I could see a log cabin that I thought would be perfect for our venue. I pulled off the Interstate and the cabin I'd seen was actually the Old McKendree Chapel; but, along the way, I spotted another large log cabin that looked unfinished and unoccupied. I stopped and asked a lady, Mary, working outside nearby what she knew about the cabin. She said, 'What do you want to know?' It turns out she was a knowledgeable Board member at Old McKendree Chapel. She relayed the history including the University's historic preservation students' former efforts adding, 'We plan to fix it up when we win the lottery, but we don't gamble.' I told Mary I was interested in saving the cabin if and when it might become available.*

"Two weeks went by, and Mary called to say they'd had a Board meeting and voted to donate the Hunter log house to us. Stunned, me and my wife, who could see the vision of the cabin, admitted to Mary that we were so busy with the wedding venue that the timing wasn't ideal. I told her if it was still available in nine months or so, it'd be more feasible. Mary called again about three weeks after and said she understood our position. She also said that with a subdivision going in nearby, they were becoming increasingly worried about the liability of keeping the cabin. I called 'my guy' whose worked on 1820s cabins before and after visiting the Hunter cabin he said there were a few questionable logs but that we could make it work. There was still the financial and time commitment, which I repeated to Mary. She smiled and said at some time in the past they'd taken up donations to have the house rebuilt but that they'd only raised $4,000. Mary then offered to give us those funds if it would help save the Hunter House. You know what? We successfully tagged, dismantled, and transported that cabin for about $800 over that budget.

"Not rushing into reconstruction ended up being advantageous. Someone contacted us after they learned we were working on a log house. They found logs behind the wall of a house they were tearing down. We ended up attracting a second, two-story cabin and after a few thousand more dollars had plenty of logs to complete our vision.

"We finally decided to incorporate the Hunter log cabin into a larger design. First, we poured a full basement. We made the Hunter dog trot's open-air breezeway wider in

Imagine Doc Mattingly's surprise in 2019 when he and his wife and daughter, Nancy and Jennifer, dined at the restaurant on the Rademaker's homestead. Doc thought something looked familiar and did a double-take to discover serendipitously that the Hunter Log House is in 2021 the Rusted Route Farms Vrbo!

order to enclose the gap by setting the full, two-story cabin in the middle. This way, Hunter's two portions flank the center nucleus. Then, to add height to the ceilings, we built 3' fieldstone wainscotting upon which we stacked the logs for taller ceilings on both levels. We ended up with three bedrooms, three full bathrooms plus a ½ bath under the stairs."[35]

Cultural Resource Surveys and Development Plans

Historic Preservation students continued their work on the Cultural Resources Survey of Historic Cape Girardeau during the 1984-1985 school year, as reported in the 10th edition of *Pastscript*. In the 18th edition of *Pastscript*, 1992-1993, it was announced that Dr. Betterly's Legal and Economic Principles of Historic Preservation class had surveyed over 200 buildings for the City of Cape Girardeau. The result was an inventory of eligible buildings for the National Register of Historic Places. "The survey will assist city planning officials to lessen potential development and demolition threats to the historic area of downtown Cape."

This is, of course, one of any number of surveys, National Register nomination applications, cultural resources surveys, development plans and other reports that SEMO HP students and grad students have initiated.

Documenting buildings is an activity that continues to this day, mostly through HP450/650: Problems in Historic Site Administration (cultural reports) and HP588: Legal and Economic Principles of Preservation (property surveys). See syllabi separately.

As of 2016, "the tally for nominations completed [by students and faculty of the HP Program] is 21 in Cape County, 1 in Butler and 2 in Scott.... The total number of properties surveyed since 1996 is 1,769."[36]

Here are some titles, as provided by Dr. Hoffman from his tenure at SEMO. These and other student project reports are conserved in the Special Collections and Archives Department in Kent Library:

1985	Water Street to Frederick Street/Bellevue to William St, plus International Shoe Factory
1991	Fountain Street to Sprigg/William Street to Morgan Oak
1993	Bellevue Street to Independence/Sprigg and Ellis Streets/Fountain Street
1994	513-531 Fountain St; 419-421 Mill; 512-550 North Middle St.
1995	Locust/Maple Neighborhood Restoration Project
1996	Sunset Area Property Survey, Part 1, W. End Blvd to Louisiana, 127 properties
1997	Sunset Area Property Survey, Part 2, Louisiana to Keller, 86 properties
1998	Sunset Area Property Survey, Part 3, Keller to Sunset Blvd, 86 properties
1999	Sunset Area Property Survey, Part 4, Sunset Blvd to Caruthers, 133 properties
2000	Sunset Area Property Survey, Part 1, RESURVEY from 1996
2000	Pacific-Sprigg Property Survey, Part 1, Morgan Oak to William, 57 properties
2000	Themis St/Morgan Oak/Water/Middle St; 600 block Good Hope St; 300 block of S Sprigg St, 170 commerical and residential properties
2001	Pacific-Sprigg Property Survey, Part 2, Morgan Oak to Themis, 64 properties
2002	Pacific-Sprigg Property Survey, Part 3, Morgan Oak to Broadway, 55 properties
2003	Pacific-Sprigg Property Survey, Part 4, Merriwether to Bellevue, 55 properties
Spring 2003	Glenn House Textiles Collection, PastPerfect Museum Software[37]
Spring 2003	Glenn House Interpretive Plan[38]
2004	Pacific-Sprigg Property Survey, Part 5, William to Themis, 32 properties
2004	Morgan Oak-Hwy 74 Property Survey, Part 1, Fountain to Sprigg, 51 properties
2005	Morgan Oak-Hwy 74 Property Survey, Part 2, Frederick to Pacific, 94 properties
Spring 2005	The Oliver House Development Plan[39]
Spring 2005	Environmental and Institutional Analysis for the Oliver House[40]

2006	Upper Broadway Commercial District Survey, W. End Blvd to Water, 117 properties
2007	Main Street Commercial District, RESURVEY from 1985, Themis to Merriwether, 39 properties
Spring 2007	The Oliver House Furnishing Plan[41]
Spring 2008	Benjamin F. Hunter Log Cabin Historic Structures Report[42]
Spring 2008	Benjamin F. Hunter Log Cabin Furnishing Plan[43]
2008	Boulevard Historic District Survey, Broadway to Highland, 75 properties
2009	Normal-Sprigg Historic District Survey, Pacific/Bellevue/Sprigg/Terrace, 86 properties
2010	North Sprigg-North Middle Local Historic District Survey, 71 properties
2011	North Middle-North Main Local Historic District Survey +supplements, 76 properties
Spring 2011	James Reynolds House Furnishing Plan[44]
Spring 2011	James Reynolds House Strategic Plan[45]
Spring 2011	James Reynolds House Structures Report[46]
2012	South Sprigg Local Historic District Survey, Part 1, 152 properties
Spring 2012	Meet Me in 1904 or, A Tale of Victorians Unbound [UI507][47]
2013	South Sprigg Local Historic District Survey, Part 2, 106 properties
Spring 2013	James Reynolds House Heritage Education Plan[48]
Spring 2013	James Reynolds House Interpretive Plan[49]
2014	South Sprigg Local Historic District Survey, Part 3, 109 properties
Spring 2015	Glenn House Heritage Education Plan[50]
Spring 2015	Glenn House Interpretive Plan[51]
2015	South Sprigg Local Historic District Survey, Part 4, 98 properties
2015	Red Star District, Part 1, area north of Sloan Creek
2016	Red Star District, Part 2, area north of Sloan Creek
2017	Red Star District, Part 3, area north of Sloan Creek
2018	Sprigg/Bertling/Big Bend Rd-Main St/Park Street

Tombstone Restoration and Cemetery Mapping Practicum

In fall 1992, Dr. Betterly's Introduction to Historic Preservation class conducted a documentation/mapping project at Old McKendree Cemetery. They recorded over 150 grave markers and prepared a site map using measurements taken from triangulating the position of each gravestone. The project assisted the Old McKendree Memorial Association to plan for the care and management of this material culture resource.

Cemetery surveying and tombstone restoration and cleaning are ongoing efforts to provide experiential learning opportunities for HP students.

Technology

"Historic Preservation majors will need expertise in graphics programs to aid in preparing site plans, building floor plans, museum layouts, cemetery mapping, and a host of other related applications. HP majors also need skill in desktop publishing and database manipulation since much of the work they will encounter in the field relies on using computers with these software capabilities. Dr. Betterly, the new [1992-1993] Director of the HP Program, plans that all HP courses use project-oriented learning experiences that take students into the field to do applied history for client. To ensure that this is possible, the department will purchase a computer station for student use to be housed in the Center for Regional History."[52]

Needless to say, technological advances since the early 1990s have grown exponentially and beyond anyone's comprehension 20 short years ago. It's a wonder where we may be technologically 20 years from now in 2041? Is there any predictive AI? Will microfilm still rule as the most economical preservation tool for 2-D materials (since all HPers know that digitization is for access and minor restoration; but, is not a preservation hallmark)? Check back as future retrospectives chronicle the unending changes.

2009 graduate Megan Atchley enjoying her internship at the Shropshire Archives in Shrewsbury, England, in 2008

Tours

Giving the St. Louis Arch a "loving" leg-up?

HP students have taken unnumbered tours of historic sites in the southeast region over the last 40 years. Some "field trips" are arranged by the HPA, others by the Center for Regional History, the Historic Preservation Program, or the Department of History and Anthropology at large. While many are day trips beyond Cape Girardeau, some have been as close to home. That is, on the SEMO campus. For instance, John O'Malley, head of Facilities Management, invited HP425 students [in 1997] to study and document the progress of the rehabilitation of the Social Sciences Building "as the building gets a new interior, but retains its old "skin."[53]

Exhibits

Mr. White's spring 1993 Museum Studies class developed an exhibit on railroad communication for the Friends of Steam Railroading in Jackson, Mo. The students developed and planned the exhibit from start to finish, including research, exhibit case design and layout, and interpretation. The exhibit, located in a caboose near the Iron Mountain Railroad station in Jackson, opened in May 1993.

Exhibits are a core component of HP student life at SEMO. Case exhibits. Panel exhibits. Even interior design story boards count, right Dr. Kellerman? Exhibit planning, design, and installation "is part of the Museum Studies practicum, which happens every other spring, alternating with the Sites practicum," according to Dr. Steven Hoffman.

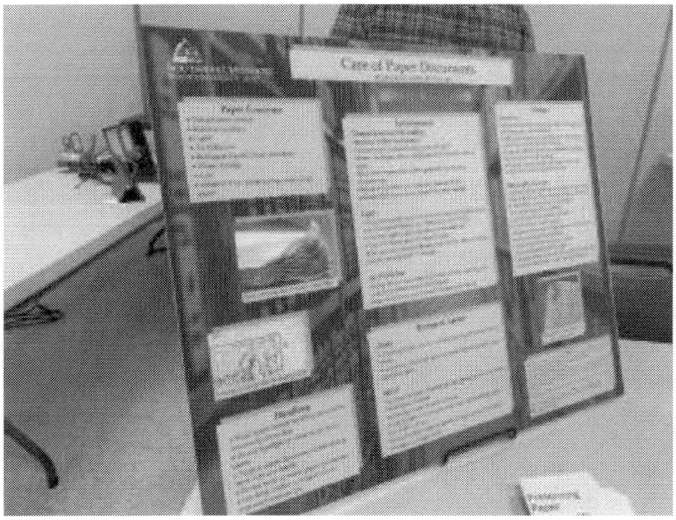

Summer Field School

"SEMO, Murray State University (KY), and the Missouri Department of Natural Resources will co-sponsor a field school in historic archaeology this summer in Ste. Genevieve, Missouri. Dr. Kit Wesler, Archaeologist, will be the chief field investigator.[54] Dr. Carol Morrow and Dr. Bonnie Stepenoff will coordinate documentary research. Students enrolled in the school will spend six weeks from June 2 through July 11 [1997], participating in an intensive research project in the material culture of colonial Ste. Genevieve. Students who wish to enroll in HP551 for six hours credit should see Dr. Morrow or Dr. Stepenoff."[55] Anne (Kern) Bauer found candid photographs "mostly of the dig site at the Delassus-Kern house south of town," of this first summer field school.[56]

This newsletter from 2000 was rediscovered in a dusty corner of the Internet:

MASTERPIECE GAZETTE Fall 2000

mostateparks.com/gazettefall2000/gazettestud.htm

Students Attempt to Solve Mysteries of the Past in Ste. Genevieve

Ten college students gathered in Ste. Genevieve to explore one of Ste. Genevieve's intriguing mysteries - the late-18th century Delassus-Kern house. Their research was part of an assignment for the Fourth Annual Historic Preservation Field School held June 13 - July 7 in Ste. Genevieve. Sponsored by Southeast Missouri State University in Cape Girardeau and the Missouri Department of Natural Resources (DNR), the field school was designed to acquaint students with a broad array of skills and techniques used in historic preservation and give them hands-on experience using them.

The field school, under the direction of Southeast Missouri State University professor, Bonnie Stepenoff, Ph.D., began with tours of the Felix Valle State Historic Site and other historic buildings in Ste. Genevieve. Hilliard Goldman, owner of the Green Tree Tavern, and Jack Luer, a historical architect, gave details of the ongoing restoration of the Green Tree Tavern and the types of research required to restore a building to its original state. They pointed out that archival documents aren't always accurate and need to be physically researched. Stepenoff says that part of the students' assignment was to "compare historical research in archives and the library to field research using close observation." A second focus of research was on African American families that resided in Ste. Genevieve.

Teams of students began unraveling the mystery of the Delassus-Kern house during the second week of the field school. Students spent the week using various methods of archival and field research to uncover facts and develop theories about the history of the house. The house served as a "laboratory" wherein students received hands-on field experience in documentation and preservation of the built environment.

As part of the Felix Valle House State Historic Site, DNR owns the Delassus-Kern house but hasn't begun restoration, partly due to the number of unanswered questions surrounding the history of the house. "Research that the class will do may help DNR with an interpretive program for the Delassus-Kern house," says Stepenoff. Jim Baker, historic site administrator at the Felix Valle House State Historic Site, feels that students are making an important contribution and is "excited about DNR's cooperation with the field school and the historic preservation program at Southeast Missouri State University. We recognize that these students are going to be in charge of preserving our cultural resources in the future."

During week three, the students conducted archeological digs under the Delassus-Kern house to try and uncover artifacts that would reveal information about the people that lived there. Larry Grantham and Ken Cole, archaeologists for DNR's Division of State Parks, supervised the digs.

Historic preservation topics were presented by guest speakers during nine lunch bag seminars held during the school. Experts in the field gave informal presentations on topics covering primary research, historic site administration, French archaeology, African American heritage, Missouri's historic sites, national landmark documentation, animal archaeology, Ste. Genevieve architecture and researching Marie LaPorte, a well-known citizen of Ste. Genevieve.

Most of the participants attending the field school were students at Southeast Missouri State University majoring in historic preservation and ranged from freshmen to graduate students. Southeast Missouri State University in Cape Girardeau is one of only four schools in the nation that offers a four-year degree program in historic preservation. Students can also earn a master's degree in history with an emphasis in historic preservation. Although the field school is not a required course in the program, it is highly recommended by Stepenoff as an "intensive field experience."

The 6th Annual Historic Preservation Summer Field School in 2002 was held in Ste. Genevieve from June 10 through July 3. Highlights under the direction of Dr. Bonnie Stepenoff, coordinator of Southeast Missouri State University's Historic Preservation

Program, were a paint analysis workshop, archaeological investigations, architectural tours, and a special lecture on the Lewis and Clark Trail.

Can you dig it? 1998 Archaeology Field School, courtesy Leslie James
Back: Matt McKnight, Stan Henderson. Center: Tiffany Timken, Leslie James, Heather Pobst, Donna Rausch, Brad Bouchard, Chad Boehlke. Front: Jeni Dockery, Mandy Wagoner, Jen Bush, Stacy Cummins, Lee Rains. Photo by Jim Baker.

Southeast Missouri State University's Historic Preservation Program conducts field studies each summer in Ste. Genevieve, Missouri, in cooperation with the Missouri Department of Natural Resources. Research areas include Missouri's French Colonial architecture, history, and culture, African American culture, and the social history of the antebellum free black community in Ste. Genevieve as it relates to the preservation and management of historic properties. The host institution for field school activities is Felix Valle House State Historic Site in Ste. Genevieve. Abstracts from the Field School's promotional literature and curriculum syllabus relay that:

Students will receive hands-on field experience in documentation and preservation of the built environment. Authentic 18th and early 19th century structures in the Felix Valle State Historic Site will serve as their learning laboratories. The goal of the Field School is to acquaint students with the broad array of skills and techniques affecting historic preservation issues and historic site management. Students will learn to gather and record architectural data, and how to effectively utilize library and archival collections in researching history, cultural landscapes and historic structures. Technical sessions will provide an introduction to preservation tools such as historic paint analysis and the use of remote sensing for examining subterranean archaeological features.

Students will be expected to participate fully in all activities, including documentation of a historic building, archaeological field work, classroom sessions, special lectures and tours of important historic houses and sites in Ste. Genevieve.

Housing for students not commuting may be arranged within the historic community. Three hours course credit may be earned at either the undergraduate or graduate level from Southeast Missouri State University. Limited scholarship support may be available.

Course Syllabus 2002
HP552: Historic Preservation Field School

I. Catalog Description and Credit Hours of Course:
Intensive field experience in documenting and preserving the built environment and the historic landscape.

II. Prerequisites:
HP100 and HP200 or equivalent or permission of instructor.

III. Purposes or Objectives of the Course:
1. To provide field experience in the documentation and preservation of the built environment.
2. To familiarize students with various methods of gathering data and understanding past cultures, including historic archaeology.
3. To deepen the student's understanding of the historic landscape.
4. To strengthen the student's skills in field observation and historical research.
5. To broaden student understanding of past human behavior from a multi-disciplinary and multi-cultural perspective.

IV. Expectations of students:
This course will be a field experience class. Students will divide their time between library and archival research and the collection of data in the field. There will also be hands-on projects relating to the study and preservation of historic buildings and objects, including work with standing structures and archaeological sites. Students are expected to: attend all sessions, read assigned materials, conduct themselves as professionals, participate fully in group projects, and complete any written assignments. Graduate students will prepare a research paper, based on original research, in consultation with the instructor(s). In general, the class will meet four days per week for four weeks during the regular summer session. Class sessions will consist of two hours of lecture/discussion and four hours of field work/lab. Lecture and discussion will focus on a specific research project relating to a historic site. Field work will include tours and inspection, photography and documentation, archaeological excavations, and hands-on demonstrations.

V. Course Outline
A. Field work methodology.
B. Research in primary and secondary sources.
C. Lectures on the built environment and the historic landscape.
D. Practice in analyzing and reporting data gathered from the field and from documentary research.
E. Field experience in techniques of preserving buildings and sites.

VI. Textbooks
Various readings will be assigned to enhance the student's understanding of observations made in the field. General readings will be assigned from the following text:

Tyler, Norman. *Historic Preservation.* New York: W.W. Norton, 1999.

Additional readings will relate to the particular site on which research is being conducted.

VII. Basis for Student Evaluation
Grades for undergraduates will be based on performance in field exercises (50%), written assignments (30%), and class participation (20%). Graduate students will be evaluated on field exercises (40%), class participation (20%), and a substantial research paper (40%).

No need to call OSHA. Students at the Field School, October 2013

Field School Seminars 2002
Ste. Genevieve, MO

Wednesday June 12:
Presenter: Dr. Eric Clements, Southeast Missouri State University
Topic: "Gambling with Your Heritage: Casinos Come to Colorado Historic Districts"
Time: Noon   Place: County Services Building

Wednesday, June 19
Presenter: Dr. Bonnie Stepenoff, Southeast Missouri State University
Topic: "Research in the Summer Field School"
Time: Noon Place: County Services Building

Field School, 2015

Wednesday June 26
Presenter: James M. Denny, Missouri Department of Natural Resources
Topic: "Lewis and Clark Trail"
Time: 4 p.m.
Place: Hotel Ste. Genevieve
Dinner: 5:30, Bolduc Room, Hotel Ste. Genevieve Cost: $10 per person
RSVP 573-883-7102

Presenter: Dr. Walter Schroeder
University of Missouri – Columbia
Topic: "Land Division in the Ste. Genevieve District, 1790-1820"
Time: 7 p.m.
Place: Hotel Ste. Genevieve
All lectures are free and open to the public!

Sponsored by:
Southeast Missouri State University, and the
Missouri Department of Natural Resources

Field School, 2015

The 8th Annual Historic Preservation Field School was conducted in Ste. Genevieve, Missouri, from June 14 through July 8, 2004. It follows suit that the 21st Annual Historic Preservation Field School was held in 2017, when Dr. Steven Hoffman posted the following:

HP552 Historic Preservation Field School (2017)

Instructors: Dr. Steven J. Hoffman
Department of History
311T Carnahan Hall
email: shoffman@semo.edu

Office Telephone: (573) 651-2808
Home Telephone: (573) 335-2820
Cell Phone: (573) 225-8393

Ms. Donna Rausch, Site Administrator
Mr. Jim Baker
Felix Vallé House State Historic Site
P.O. Box 89 Ste. Genevieve, MO
email: felix.valle.state.historic.site@dnr.mo.gov
course webpage: cstl-cla.semo.edu/shoffman/hp860

Office Telephone: (573) 883-7102

Course Description:
A workshop for graduate, undergraduate and continuing education students involving intensive field experience in documenting, interpreting, and preserving the built environment and the historic landscape. Students will receive hands-on field experience in documenting, preserving, and interpreting the built environment and the historic landscape, including work with standing structures and historic sites. Authentic 18th and early 19th century buildings in the Felix Valle State Historic Site will serve as learning laboratories, providing hands-on investigation of Ste. Genevieve's rare vertical log structures. The goal of the Field School is to acquaint students with the broad array of skills and techniques affecting historic preservation issues and historic site management. Students will learn to gather and record architectural data, utilizing library and archival collections in researching history as tools for understanding historic structures. (3 credit hours)

Objectives of the Course:

1. To provide field experience in the documentation and preservation of the built environment.
2. To familiarize students with various methods of gathering data and understanding past cultures, including material culture and historic archaeology.
3. To deepen students' understanding of the historic landscape.
4. To strengthen students' skills in field observation and historical research.
5. To develop a basic understanding of the operation of historic sites.
6. To broaden student understanding of past human behavior from a multi-disciplinary and multi-cultural perspective.

Student Learning Objectives:

1. Students will be able to demonstrate knowledge and skills relevant to the documentation of historic buildings, landscapes, or sites.
2. Students will be able to demonstrate knowledge and skills relevant to the preservation of historic buildings, landscapes, or sites.
3. Students will be able to demonstrate knowledge and skills relevant to the interpretation of historic buildings, landscapes, or sites.

Expectations of students:

Students are expected to: attend all sessions, read assigned materials, participate in all field trips and field studies, conduct themselves as professionals, participate fully in group projects, and complete any written assignments. In addition, graduate students will have additional readings and be responsible for leading portions of discussion.

Proposed Schedule: Meet daily, 9:00 a.m. – 5:00 p.m. (Wednesday, May 24: 9-5, 6:00-9:00) at the Creole House, 339 St. Mary's Road, Ste. Genevieve, Missouri.

Monday, May 15, 2017
Introduction and Orientation
Planning Process for Historic Sites Presentation of Interpretive Plan
Tour: Bauvais-Amoureux House
Felix Vallé House State Historic Site

Tuesday, May 16
Assignment for Tuesday:
Read Interpretative Plan; Read Invitation to Vernacular Architecture, Introduction and Chapter 1 (pp. xiii-18)
Meet at Creole House
Tour: Bolduc House
Ste. Genevieve Interpretative Center
Guibourd-Vallé House
Ste. Genevieve Memorial Cemetery
Ste. Genevieve Museum
Discuss Felix Valle interpretative plan

Field School, 2013

Monday, May 22
Assignment for Monday: Read Invitation to Vernacular Architecture, Chapter 2 (pp. 19-43) Measured Drawings (Richard Young, famous Southeast HP graduate) (Location TBA)

Tuesday, May 23
Assignment for Tuesday: Read Invitation to Vernacular Architecture, Chapter 3 (pp. 45-62)
Document Transcription
Green Tree Tavern tour Lunch (11:00-12:00)
Photography Workshop (12:00-4:00) (Location TBA)
Work on Projects (time permitting)

Wednesday, May 24
Assignment for Wednesday: Read Invitation to Vernacular Architecture, Chapter 4 (pp. 65-81)
Timber framing workshop, Durand Cabin (Jesse Francis, St. Louis County Parks Department and famous Southeast HP graduate)

Evening event: Creole House – Screening and discussion of Under These Same Stars: The Céladon Affair, a film about an episode in Ste. Genevieve's history.

Thursday, May 25
Archeology (Location TBA)

Friday, May 26
Window Repair Workshop
Heritage Education project: Paper Marbling

Tuesday, May 30
Assignment for Thursday: Work on Projects
Field Assignment: Find 2-3 variations of a mortise and tenon joint at the Guibourd-Vallé house and sketch them. Time: TBA
Norman Truss races. Time: TBA
"What the Tour Guides Won't Tell You" walking tour

Field School, 2016

Wednesday, May 31
Groups Present Interpretive Plans (Green Tree Interpretative Plan)
Individuals present transcribed documents
Tour Shaw House kitchen to evaluate completed projects
BBQ lunch – Lion's Club Park, Market Street

Textbooks

Thomas Carter, Elizabeth Collins Cromley, *Invitation to Vernacular Architecture: A Guide to the Study of Ordinary Buildings and Landscapes* (University of Tennessee Press, October 24, 2005) Various additional readings will be assigned to enhance student understanding of observations made in the field and to relate to the particular projects being pursued.

Basis for Student Evaluation

Grades for undergraduates will be based on performance in field exercises (50%), written assignments (30%), and class participation (20%). Graduate students will be evaluated on field exercises (50%), written assignments (20%), and class participation (30%), including leading portions of discussion of additional readings.

MORE CURRENT CLASS SYLLIBI IN CHAPTER 5.

Field School, 2021

Field School, 2021

Internships

One flagship identifier of the SEMO HP Program, other than its attraction to prospective students across the United States, is its internship program. The HP Program was the very first degree program at SEMO to institute an internship as part of required scholarship towards graduation.

1996 graduate Scottie Cross, ca. 2020

Interns have opportunities to work in archives, museums, historic sites and preservation organizations and agencies, depending on their specific area of interest. It is usually students' first chance to gain practical experience in the "real world."

Internships last from 8-12 weeks consisting of 400 hours for 9 credit hours of upper-class curriculum. More than glorified volunteer positions, internships provide students the unprecedented opportunity to participate in—and often lead or direct—the day-to-day operations and activities of historical and cultural heritage sites. Further, student interns work under the supervision of professionals in the field . . . curators, archivists, architectural historians, etc., where they receive real life lessons in the context of working with actual two and three-dimensional artifact collections. It affords students the chance to exercise what they have learned in the classroom setting, applying theory to on ground, or real life situations. Some internships offer a stipend and even housing; many do not.

The following Q&A was developed over time, a version of which was posted online in 2001-2002 during Dr. Stepenoff's tenure. Long "lost," it was located and transcribed from the "Way Back Machine" at archive.org in 2021:

<u>Who must complete an internship?</u>

All historic preservation majors are required to complete an internship in order to qualify for the bachelor of science in historic preservation.

Master's degree candidates in history with a preservation emphasis are encouraged to complete an internship.

Rach Teasdale, 2020

When will I do my internship?

Most students complete the internship during the summer between their junior and senior years. But other timetables are acceptable. Before starting an internship, students must complete the following courses: HP100, HP200, and one of the following sequences: HP405 and HP407, HP410 and HP417, or HP 420 and HP425.

HP student Rachel Kincy's Internship, 2021

In other words, before beginning an internship, students must have completed a 400-level problems course in historic preservation.

Graduate students must have completed sixteen hours of graduate course work.

How much credit do I receive for my internship?

Undergraduate internships are worth nine hours of credit toward the degree. Graduate internships are worth three hours of credit toward the degree.

How do I enroll for my internship?

Historic preservation internships are listed as courses numbered HP500-501-502. Graduate internships are listed as course number HP640. In each case, you must obtain permission to enroll from the internship coordinator.

What must I do to receive a grade and credit for my internship?

Each internship must involve at least 400 hours of work at an appropriate facility. The work done must be sufficiently varied and challenging to provide a significant learning experience. Students should inform the internship coordinator of the plan of work. The internship coordinator must make contact with the supervisor of the work, in order to insure that internships are a valuable experience for the student. In order to receive a grade for the internship, students must complete the following:

HP student Allyson Ingli's Internship, 2021

- a concise daily journal of activities during the internship
- a five-page typewritten report on a special project involving significant research or a professional level of activity*
- a 30-minute oral presentation on the intern's experiences (to be presented to an audience of historic preservation students) OR a twenty-page paper describing the internship experience

* The student and the internship coordinator should agree in advance on a special project.

Each graduate internship must involve at least 150 hours, with all the qualifications above but with a more "substantial report."

<u>Where can I do my internship?</u>

Students should take the initiative in arranging an internship at a museum, archival facility, historic site, or historical agency. The internship coordinator maintains files on these and other facilities. Set your sights high. Explore the possibilities. Opportunities abound.

^ HP student Peter Reckling's Internship, 2021
v HP student Jared Anderson's Internship, 2021

Historic Preservation internships have included partnerships with the following institutions since 1983.

This list of nearly 200 sites is nearly but not all inclusive, and has been compiled from back issues of *Pastscripts*, from Dr. Steven Hoffman, and 2021-22 HPA VP Allyson Ingli's input, and from an informal survey on both the HPA and HP Alumni Chapter Facebook pages in June 2021.

Graduates are encouraged to contact their alumni chapter to improve this impressive recap:

SEMO HP STUDENT INTERNSHIPS, 1981-2021

Institution	Location	Date/s
Abraham Lincoln Presidential Library and Museum	Springfield, IL	2015
Adirondack Architectural Heritage	Keeseville, NY	2017
Alabama Historical Commission	Montgomery, AL	2019
American Czech Educational Center	St. Louis, MO	2006
Andersonville National Historic Site	Americus, GA	2006
Arizona Capital Museum	Phoenix, AZ	2012
Arizona State Library, Archives and Public Records	Phoenix, AZ	2012
Arkansas State Historic Preservation Office	Little Rock, AR	1996; 2015
Arthur Bowring Sandhills Ranch State Historic Park	Merriman, NE	1991
Association of Midwest Museums	St. Louis, MO	2007
Battle of Pilot Knob State Historic Site	Pilot Knob, MO	
Battle of Westport Visitor Center and Museum	Kansas City, MO	2016
Baxter Spring Heritage Center and Museum	Baxter Springs, KS	2004
Belleville Main Street	Belleville, IL	2006
Bishop Grosseteste University	Lincolnshire, UK	2012
Blue Ridge Parkway (National Park Service)	Asheville, NC	
Bolduc House Museum	Ste. Genevieve, MO	2010; 2017
Bollinger Mill State Historic Site	Burfordville, MO	2008; 2011; 2014; 2015; 2016; 2017
Bureau of Reclamation	Denver, CO	2005
B-W Consultants, Inc.	Springfield, IL	2009
Campbell House Museum	St. Louis, MO	2017
Cape Girardeau County Archive Center	Jackson, MO	2003; 2009; 2015; 2016; 2019; 2021
Cape River Heritage Museum	Cape Girardeau, MO	2010
Carondelet Historical Society	St. Louis, MO	2019
Castine Historical Society	Castine, ME	2016; 2017
Center for Historic Preservation Ball State University	Muncie, IN	2006
Chateau de Mores	Medora, ND	2010
Chemical Weapons Museum	Fort Leonard Wood, MO	2013
Chillicothe Main Street	Chillicothe, MO	2015; 2021
City of Cape Girardeau-Planning Department	Cape Girardeau, MO	2001
City of Fredericktown	Fredericktown, MO	2015
City of Jackson	Jackson, MO	2011; 2016
City of Sedalia	Sedalia, MO	2013
Civil War Trust	Washington, D.C.	1994-1995
Coldwell Banker Platinum Partners	Savannah, GA	2011
Concordia Theological Seminary Fort Wayne, Archives	Fort Wayne, IN	2021
Conner Prairie	Fishers, IN	1994-1995; 2013
David Davis Mansion	Bloomington, IL	2006
Delta Cultural Center	Helena, AR	2012
Delta Gateway Museum	Blytheville, AR	2013; 2014
Deutschheim State Historic Site	Hermann, MO	2016

Downtown Washington, Inc.	Washington, MO	2006; 2009; 2012; 2013; 2014; 2015; 2016; 2019
Drake Well Museum and Park	Titusville, PA	Fall 1991
DuPage County Historical Museum	Wheaton, IL	2015
Early American Museum	Mahomet, IL	2007
Eisenhower National Historic Site	Gettysburg, PA	2019
Eugene Field House & St. Louis Toy Museum	St. Louis, MO	2000; 2005; 2009; 2010; 2013; 2015; 2017
Federal Surveys	Atlanta, GA	1994-1995
Felix Vallé House State Historic Site	Ste. Genevieve, MO	2000; 2005; 2012; 2015; 2016; 2017; 2019
Fort Donelson National Battlefield	Dover, TN	2016
Fort Ticonderoga	Ticonderoga, NY	2007
Franklin County Planning and Zoning	Union, MO	2011
Ft. Davidson State Historic Site	see Battle of Pilot Knob State Historic Site	
Ft. Larned National Historic Site	Ft. Larned, KS	1995; 1998
G.W. Blunt White Library Mystic Seaport	Mystic, CT	2016
General Alfred M. Gray Marine Corps Research Center,	Quantico, VA	2007
George Rogers Clark National Historical Park	Vincennes, IN	2012
Gettysburg National Military Park	Gettysburg, PA	1998; 2014
Gettysburg National Military Park	Gettysburg, PA	2014
Main Street Chillicothe	Chillicothe, MO	2014
Glenn A. Black Laboratory of Archaeology	Bloomington, IN	2010
Glenn House	Cape Girardeau, MO	1983
Hancock Shaker Village	Pittsfield, MA	2017
Harry S Truman National Historic Site (home)	Independence, MO	1983-1984
Harry S Truman Presidential Library and Museum	Independence, MO	1992; 2010
Historic Macon Foundation	Macon, GA	2016
Historic Quincy Business District	Quincy, IL	2011
Historic Travellers Rest Historic House Museum	Nashville, TN	1996
History and National Historic Landmarks Program (NPS),	Omaha, NE	2008
History Nebraska (SHPO)	Lincoln, NE	2021
Hunter-Dawson Historic Site	New Madrid, MO	1983-1984
Hunter-Dawson State Historic Site	New Madrid, MO	2013; 2021
Idaho State Historic Preservation Office		
Illinois Historic Preservation Agency	Springfield, IL	2005
Indiana Governor's Office-Indiana Main Street	Indianapolis, IN	1993
Indiana Landmarks	Indianapolis, IN	1988; 1994
Iowa Main Street	Des Moines, IA	2013
Oklahoma History Center	Oklahoma City, OK	2013
Missouri Main Street	Branson, MO	2014
Jackson County (Mo.) Historical Society Archives	Independence, MO	2013
Jamestown-Yorktown Foundation	Williamsburg, VA	2008
Jefferson Barracks County Park	St. Louis, MO	
Jefferson Barracks Heritage Foundation	St Louis, MO	2007; 2009
Jefferson National Expansion Memorial	St. Louis, MO	2005
Jewish Museum Milwaukee	Milwaukee, WI	2015

Jimmy Carter National Historic Site	Americus, GA	
Kansas City Museum	Kansas City, MO	2016
Karen Bode Baxter, Historic Preservation Consulting	St. Louis, MO	2015
Knife River Indian Villages	Stanton, ND	2010
KSK Architects, Planners, Historians Inc.	Philadelphia, PA	2003
Lafser & Associates	Jackson, MO	2006
Lake Wappapello	Wappapello, MO	2005
Landmark Society of Western NY	Rochester, NY	2017
Library of Congress	Alexandria, VA	2012
Liturgical Consulting and Design	Chicago, IL	2012
Luther Center	Wittenberg, Germany	2006
Lutheran Heritage Center & Museum	Altenburg, MO	2006; 2013; 2020
Mackinac State Historic Parks	Mackinaw City, MI	2015
Macon County History Museum	Decatur, IL	2017
Main Street Blytheville	Blytheville, AR	2013
Main Street Chillicothe	Chillicothe, MO	2011; 2012; 2015
Main Street Mobile	Mobile, AL	2005
Maine Preservation	Yarmouth, ME	2017
Mastodon State Historic Site	Imperial, MO	2006; 2008; 2015
McFaddin-Ward House	Beaumont, TX	2014
McHenry County Historical Society	Union, Illinois	2008
Minnesota Historical Society	St. Paul, MN	2006
Missouri Botanical Garden	St. Louis, MO	Dec 1996
Missouri Civil War Museum	St. Louis, MO	2015; 2016; 2017; 2021
Missouri Department of Transportation	Jefferson City, MO	2001; 2013
Missouri Historical Society	St. Louis, MO	1983-1984; 1988; 1994-1995; 2007
Missouri Main Street Connection	Branson, MO	2011
Missouri Preservation	Columbia, MO	2012
Missouri State Archives, Local Records Division	Jefferson City, MO	1994; 2021
Missouri State Historic Preservation Office (SHPO)	Jefferson, City, MO	1999
Missouri State Museum and Jefferson Landing State Historic Site	Jefferson City, MO	2006; 2007; 2008
Missouri State Museum	Jefferson City, MO	1994-1995; 2014
Missouri State Society Daughters of the American Revolution	Dearborn, MO	2011
Montana State Museum	Helena, MT	
Mt. Hood National Forest	Sandy, OR	2012
Museum of Missouri Military History	Jefferson City, MO	2019
Museum of Osteopathic Medicine, A.T. Still University	Kirksville, MO	2013
National Archives and Records Administration	Denver, CO	1996
National Archives and Records Administration	Kansas City, MO	
National Archives and Records Administration	St. Louis, MO	2014; 2015; 2019
National Churchill Museum	Fulton, MO	2010
National Museum of Transportation	St Louis, MO	1983-1984; 2009; 2010
National Naval Aviation Museum	Pensacola, FL	2017
National Park Service Curatorial Center	Landover, MD	2001
National Park Service HP Training Center	Frederick, MD	2000
National Park Service Regional Office	Omaha, NE	2008

National Park Service	Anchorage, AK	2012
National Railroad Museum	Green Bay, WI	2012
National Trust for Historic Preservation	Washington, DC	1997; 2017
National World War II Museum	New Orleans, LA	2017
Navy Museum	Washington, D.C.	
New Jersey State Historic Preservation Office (SHPO), Trenton, NJ		1999
New Orleans Jazz Museum	New Orleans, LA	2021
Newton City Hall	Newton Centre, MA	2011
Newton Planning Department	Newton, MA	2011
Nodaway Valley Historical Museum and Clarinda Carnegie Art Museum	Clarinda, IA	2017
North Carolina Maritime Museum	Beaufort, NC	2013
Old Deerfield Village		1981
Old Town Cape, Inc.	Cape Girardeau, MO	2012
Olympia Historic Preservation Office	Olympia, WA	1994-1995
Omaha Historic Preservation	Omaha, NE	1994-1995
Oneida Community Mansion House	Oneida, NY	2017
Open Palace Programme	Lincoln, England	2015
Oshkosh Public Museum	Oshkosh, WI	
Ozark National Scenic Riverways	Van Buren, MO	1994-1995; 1997
Paine Art Center and Arboretum	Oshkosh, WI	1994
Papers of Abraham Lincoln	Springfield, IL	2015
Pennsylvania Downtown Center	Harrisburg, PA	2006
Pennsylvania Historical and Museum Commission	Harrisburg, PA	2006
Pennyroyal Area Museum	Hopkinsville, KY	2012
Perry County Historical Society	Perryville, MO	2019
Phelps County Historical Society	Rolla, MO	2014; 2017
Pierre Office of History	Pierre, SD	Jan 1997
Pink Palace Museum	Memphis, TN	2017
Preserve Oregon	Portland, OR	2015
Preserve Rhode Island	Providence, RI	2010
San Francisco History Center, San Francisco Public Library, San Francisco, CA		2008
Saxon Lutheran Memorial	Frohna, MO	2020
Scott Joplin House State Historic Site	St. Louis, MO	2006
Shawnee National Forest	North Vienna, IL	2013; 2015
Shiloh National Military Park	Shiloh, TN	1994-1995; 1997; 2010
Shropshire Archives	Shrewsbury, England	2008
Smithsonian American Art Museum	Washington, DC	2009
Soldiers' Memorial	St. Louis, MO	2006; 2021
South Dakota State Historical Society (SHPO)	Pierre, SD	2006
South Union Shaker Village	Auburn, KY	2012
St. Charles County Historical Society	St. Charles, MO	2017
St. Charles County Park	St. Charles, MO	2010
St. Clair County Historical Society	Belleville, IL	2017
St. Joseph Museums, Inc.	Joseph, Missouri	2012
St. Louis County Parks	St. Louis County, MO	1994-1995
St. Louis Cultural Resources Office	St. Louis, MO	2000
St. Louis Science Center	St. Louis, MO	2016
Stars & Stripes Museum / Library Association, Inc., Bloomfield, MO		2007; 2019

State Historical Society of Iowa	Des Moines, IA	2019
Sturbridge Village	Sturbridge, MA	2017
Tennessee Historical Commission	Nashville, TN	2011
Tennessee Main Street Program	Nashville, TN	2011
Tower Grove Park	St. Louis, MO	Dec 1996
Trail of Tears State Park	Jackson, MO	1988; 2007
U.S. Army Chemical Corps Museum	Fort Leonard Wood, MO	2005
Ulysses S. Grant National Historical Site	St. Louis, MO	2011; 2021
US Army Corps of Engineers Mandatory Center of Expertise	St. Louis, MO	2007; 2011
US Army, Cultural Resource Management	Fort Wainwright, AK	2011
Vermilionville Historic Village	Lafayette, LA	1992
Vicksburg National Military Park	Vicksburg, MS	
Volkening Heritage Farm	Schaumburg, IL	2007
Wabash County Historical Museum	Wabash, IN	2006
Washington University Libraries	St. Louis, MO	2013
Watkins Woolen Mill	Liberty, MO	1983-1984
Whipple Company Store and Appalachian Heritage Museum	Scarbro, WV	2012
Wichita-Sedgwick County Metropolitan Planning Department	Wichita, KS	2006
Wickliffe Mounds	Wickliffe, KY	
Wilmington Railroad Museum	Wilmington, NC	1995
YMCA of the Rockies	Estes Park, CO	2015

A June 2021 Facebook survey asked students and former graduates where they conducted their internship. The post reached 749 people and 267 engagements, according to Facebook. Some made the following comments on their internship experience:

"A SHPO is an excellent opportunity to use skills learned in class and to learn a great deal -- and not just about what you are doing, but what others are doing." Sara Andre (BSHP, 1999), NJ SHPO

"She killed herself in your room." Stacy Reaves, Ph.D. (BSHP, 1995), Fort Larned NHS

"Words cannot express the wonderful experiences I had during this internship. The knowledge gained is invaluable." Andrew Halter (BSHP, 2000), MO SHPO

"This is the exact courtroom that the Miranda rights trial was held in. The most memorable part about my internship was finally getting to put all my book learning into practical experience. I did an entire silver collection from adding to the collection, cleaning it up, adding to the inventory list, and creating a custom storage box for the all the different pieces. A memorable quote that my boss liked to tell me was "learn something every day so that you can be that much better at your job tomorrow."
Reid McCutcheon (BSHP, 2013), Arizona Capital Museum, Phoenix, AZ, 2012

"Hot, dry, and life-changing." Creager Smith (BSHP, 1989), LaPorte County, IN

"I experienced a real museum and was shocked that it wasn't run the way we learned in our studies. I learned about real life, people, and how to adapt."
Bridgette Kelch (BSHP, 2000), Eugene Field House & Toy Museum, Summer 2000

"I remember being amazed that I could step in and process a major archival collection relating to Harry Truman's administration without any guidance, and I knew EXACTLY what to do...and did it....and a LOT MORE. I felt a valuable asset to the staff during that fantastically rewarding summer. I know people who lived through the Great Depression (my grandparents) and they KEPT EVERYTHING. Mr. and Mrs. Truman were the same. But it was funny to see an entire shoe box full of old shoestrings PROPERLY preserved from their home. That prepared me for just about everything later in my career. Ha! Except accepting a donation of a local veterinarian's materials from the 1880s that included—when we got around to inventorying everything—such artifacts as a vial of Anthrax and a walking cane made out of a prized bull's 3 ft. penis. Ah dios mio!" David W. Jackson (BSHP, 1993), Harry S Truman Presidential Library and Museum, 1992

"I look back fondly at my memories of Shiloh, from ranger programs and cleaning the National Cemetery to bronzing the UDC monument and documenting the myriad of markers throughout the park; all of it immersive and immeasurable to further the foundation of my HP experienced at SEMO. My love of research led me to doing something everyone said was impossible - I researched the history of the Iowa monuments and wrote to all of the Iowa legislators, reminding them of the sacrifices of her native sons and of the monuments erected to their memory and asked for money to repair the Iowa monuments at Shiloh. And the money was allocated, and the monuments restored. The pen in Fame's hand is once more poised to "inscribe their names on the enduring rock." Mandy K. Ford (BSHP, 1999), Shiloh National Military Park, 1997

"*The internship with South Dakota really changed my Outlook. I was looking for museum employment prior to interning at the SHPO. Had I not had that experience I would not be working in the Arkansas SHPO currently. The supervisor at the SHPO office was a SEMO HP Grad and it made the experience even better."* Travis Ratermann (BSHP 2007), SD SHPO

"*Best internship ever.*" Kim Proctor (BSHP 2013), MoDOT, 2013

"The most memorable experience at the Missouri State Archives was getting to lunch with Missouri's First Lady at the Governor's Mansion. Also, being a contributor to the Dictionary of Missouri Biography was an honor." Misty Mabuce (BSHP, 1995)

"I was helping to reorganize a storage room when I came across a box that had grenades. I tell my boss and she immediately goes, "Put the box down slowly." Someone came to pick them up to figure out if they were active or not and luckily, they had holes drilled in them. But for a second we thought there were live grenades in the museum." Dan Will (BSHP, 2016), DuPage County Historical Museum, 2015

"The big exhibit that summer was on the 1904 World's Fair. I helped clean the red marker from a plaster lion's eyes using 0000 steel wool. That lion is still on exhibit at the museum over 30 years later!!"
Susan Haake (BSHP, 1989), Missouri Historical Society, Forest Park, St. Louis, 1988

"For me, it confirmed that I wanted to focus my preservation career towards working in the Main Street program which was relatively new at the time. My internship demonstrated there was a viable career path that blended my interest in economic/community development to Historic Preservation. Since I was the first from the SEMO HP program to venture into the Main Street Program, my experience from my internship and beyond allowed me to connect future grads of the program and serve as a resource and contact for them as well in the field." John Bry (BSHP, 1995), Governor's Office Indiana, Indianapolis-assigned to Department of Commerce and Indiana Main Street, 1993

"The experience I gained during my internship (survey of NHLs, preparation of NHL newsletter, final review and packaging of HABS documentation, building partnerships, working within a federal agency, and even assembling a model of an NHL) directly prepared me for my role now as architectural historian at Tennessee Valley Authority. As a native of Cape County, it also taught me that I could be alone away from home and still be successful. I also learned the importance of networking with fellow SEMO grads. There were several in the same office, including Stephen Rogers, and I occupied a cubicle previously inhabited by the infamous Brian Keith McCutchen! I answered a lot of phone calls for him!"

Hallie Fieser Hearnes (BSHP, 2009), History and National Historic Landmarks Program, Midwest Regional Office of the National Park Service, Omaha, Nebraska, 2008

To which Brian McCutchen replied: *"Infamous" is correct! I did not know that you had the coveted window view. I loved watching the Mo River and construction of the pedestrian bridge."*

"Though already having had a passion for battlefields, monuments/cemeteries, and related resources, I had never before imagined that I would be lined up with an internship that included those very resources of interest – and more. My summer not only provided experiences with hands-on preservation activities, it broadened my view of what a career in the historic preservation field could be. And it was exciting! Twenty-nine years and multiple positions and places with the NPS later, I often reflect upon and share about how an HP internship is more than ten weeks of learning, but, hopefully, something that one finds exciting, eye opening, and perhaps the influence for, or start of a career in something that you love." Brian Keith McCutchen (BSHP 1992; MAHP 1995), Shiloh National Military Park, Tennessee, Summer 1992

"I not only learned how to cook bread on a wood-burning stove, milk a cow, make butter and create a number of German American foods using basic kitchen tools now (if the apocalypse comes, beep me); I also was given full creative license for an interpretive exhibit, re-did their entire volunteer manual (which barely existed) and learned a lot about event planning and management. This was quite possibly one of the best summers of my life. This also taught me that most people have no idea the great resources they have in their own community. This site was in the middle of one of the biggest suburbs in the Chicago-land area and almost every visitor had no idea it was right around the corner from their local IKEA."

Alyssa Phares-Fee, Volkening Heritage Farm, 2007

"It was the craziest risk of my life to spend three months in England, but I touched things that were older than the founding at Jamestown. I experienced so many unique things in that internship that it was totally worth it to go overseas!" Megan Atchley, Shropshire Archives, Shrewsbury England UK, 2008

"We primarily worked in the Amoureux House, doing tours for visitors. Highlights are building a diorama of the house's poteaux-en-terre construction and participating in the town's summertime festivals."

Danielle Lewon, Felix Valle House State Historic Site and the Amoureux House, 1995

How do I apply for an internship?

The first step is to write a letter of interest to the agency offering the internship. You may also write letters to agencies, asking if they offer internships. In these letters, identify yourself, explain your intentions, and briefly state your qualifications. You may enclose a resume, or you may offer to send a resume later.

Prepare a resume that highlights your field experience, academic training, and special interest in preservation. If you need assistance, ask an instructor or the internship coordinator. You may also ask for assistance from the office of Career Services on campus.

Dr. Eric Clements has advised the Historic Preservation Association for 20 years

If the agency expresses interest in your application, submit all forms and materials they request. Make telephone contact with someone in the agency. Inform the internship coordinator of your progress. When negotiations become serious, the coordinator will contact the agency to explain our internship program.

Please be sure to be courteous and businesslike in dealing with the agency. When in doubt, talk to the coordinator.

Will I be paid for my internship?

Many students receive stipends, wages, or some sort of financial support from the facilities at which they do their internship. Arrangements must be made directly with the facility. Paid internships in historic preservation may also available through the National Council for Preservation Education (NCPE), which offers a variety of internships on a competitive basis every year. Students must apply for these internships. You may obtain application forms an information from the internship coordinator.

Is there anything else I should know?

While engaged in the internship, the student should maintain a cheerfully professional attitude. Interns should report to work promptly, fulfill all assignments to the best of their ability, discuss problems with supervisors and co-workers in a calm and constructive manner, and contact the internship coordinator if problems become serious. Whether paid or unpaid, an internship is a job. Interns should honor all commitments to the facilities that employ them.

The internship coordinator will be more than happy to assist in submitting applications, preparing resumes, and setting up internships. Please feel free to call the HP Coordinator.

SEMO HP Alumni and Mattingly Award Winners with Dr. Mattingly and Dr. Hoffman
35th Annual HPA Banquet, 2016

5

Curriculum Steeped in Conservation

Archivists in SEMO's Special Collections and Archives Department in Kent Library have extracted the course descriptions of HP classes from official SEMO Bulletins, year-by-year, starting in 1981. These will be assembled in the SEMO HP Alumni Records Collection in University Archives, and also on the member's only section of the alumni chapter's website (semohpalumni.org).

"The Way Back Machine" (archive.org) also provided snapshots of some core HP curriculum from the 2001-2002 and 2002-2003 academic years. Here are two samples from nearly 20 years ago, still clinging to the WWW as of 2021.

David W. Jackson's SEMO HP coursework, 1991-1993, donated to the Special Collections & Archives Department in Kent Library

Course Syllabus
HP200: Techniques of Research in Local History
Southeast Missouri State University, Spring 2002

Instructor: Dr. Bonnie Stepenoff, Department of History — Telephone: 651-2831
305 Social Science Building
Office Hours: Mon. 1-3, Tues. 1-3, and by appointment — E-mail: bstepenoff@semo.edu

This course abides by university policies regarding attendance and academic honesty.

I. Catalog Description and Credit Hours of Course:
Introduction to the value, techniques, and methods of research in local history for the preservationist. (3 credit hours)

II. Prerequisite(s): None.

III. Purposes or Objectives of the Course:
This course is designed to introduce students to methods of research in local history, placing local history in the context of national and international issues and themes.

During the semester the student in this course should:

1. Gain skills in finding and using secondary and primary source material for research.
2. Gain skills in evaluating and analyzing historical research materials.
3. Gain skills in presenting the results of research in oral and written form.
4. Gain an appreciation of local history, its value, and significance in understanding broader historical themes.

IV. Expectations of Students:
Students are expected to complete assigned readings, participate in discussions and field activities, and complete all assignments. Students will be expected to keep field notes, documenting research activities. Students will be expected to submit three short research reports and contribute substantially to the preparation of a final research report. Students will be expected to give informal oral reports on their individual research. Class participation will be a factor in the final grade. There will be two examinations.

Graduate students enrolled in the course will complete a special project to be defined during discussions with the instructor.

V. Course Content or Outline (Indicate number of class hours per unit or section):
Researching individuals (4 weeks)
Researching communities (7 weeks)
Analyzing and presenting results of research (5 weeks)

VI. Textbook(s) and/or Other Required Materials or Equipment:
Loretto Dennis Szucs and Sandra Hargreaves Luebkin, eds., *The Source* (Salt Lake City, Utah: Ancestry Incorporated, 1997).

VII. Basis for Student Evaluation

Students will be evaluated on class participation (including participation field exercises), performance on two examinations, and the quality of one oral, three written reports, and a final report (to be prepared by the class as a group).

Numerical Grading System	
First exam 40	270+ = A
Second exam 40	240+ = B
Field notes 30	210+ = C
Research reports (3) 90	180+ = D
Contribution to final report 60	
Participation 40	
TOTAL 300	

Tentative Class Schedule

Researching Individuals

Jan. 14-Feb. 15 Each student will conduct research on an individual who made a significant contribution to U.S. railroad history. A list of names will be provided. Classroom activities and reading assignments will guide students in locating and using biographical reference tools and source materials. Students will submit a brief (3-5 page) biography, **fully documented**, with a guide to available source materials. Students will also submit research notes and copies of material used in research. The report will be due February 15. Reading assignment: *The Source*, Chapter 2, 3, 9, 10 and 11. The textbook is to be used as a reference book, as needed. First examination: February 15.

Researching Communities

Feb. 18- Mar. 8 The class will research a particular community (downtown Cape Girardeau) using a variety of sources. Each student will submit a brief (3-5 page) report, identifying and describing the image in a historical photograph relating to the community. Students will work in teams of two or three, but each student will submit an individual written report on the group's findings. Groups will report orally on their findings. Each student will also submit research notes. The report will be due on March 8. Reading assignment: *The Source*, Chapter 1.

Spring Break -- Mar. 18-24

Mar. 11 - April 5 The class will continue research on the local community, attempting to reconstruct the community's past and relate it to the broader historical context. Each student will intensively research one local individual, family, or historic property connected with the development of railroads in Cape Girardeau, Missouri. Students will submit a brief summary of findings and research notes, including copies of three articles from historic newspapers. Reading assignment: *The Source*, Chapters 4, 5, 6, 7, 8, 12. Second Examination: April 5.

Analyzing and Presenting the Results of Research

Apr. 8-May 3 The class will write a history of railroading in Cape Girardeau. Each student will submit a brief (3-5 page) well-written essay (or chapter) to be included in the history, which will be compiled, edited, copied, and spiral-bound. It is essential that students attend classes and work with the group to complete this project. Each student will prepare material to be included in the report, submitting drafts for review. Panels of students will work together to discuss and revise drafts. Final compiled and edited text due on May 3.

Course Syllabus
HP405: Archives and Special Collections Management
Southeast Missouri State University, Spring 2003

Instructor: Dr. Bonnie Stepenoff
Dr. Stepenoff is coordinator of the Historic Preservation Program.
Office: Department of History, 305 Carnahan
Telephone: 651-2831
E-mail: bstepenoff@semo.edu
Office hours: Tuesday 1-3, Wednesday 3-4, or by appointment
Class will meet MWF 2 p.m. in Carnahan 109.

Here are two useful links for the course:
Southeast Missouri State University Special Collections and Archives
National Archives and Records Administration

This course abides by university policies regarding attendance and academic honesty.

I. Catalog Description and Credit Hours of Course:
Study of the philosophy, methods, and techniques of preserving and managing archives and historical collections. (3 credit hours)

II. Prerequisite(s): None.

III. Purposes or Objectives of the Course:
This course will provide a comprehensive introduction to the preservation, use, and management of historical documents and other special collections. Upon successful completion of the course, the student will have a general understanding of the problems and challenges faced by professionals charged with preserving our documentary heritage.
Primary objectives:
1. To introduce the student to the management, care and use of archival materials.
2. To introduce the student to fundamental principles and methods of information storage and retrieval.
3. To deepen the student's understanding of the importance of our documentary heritage.

IV. Expectations of Students:
Students are expected to read assigned materials, participate in class discussions, and take an active part in hands-on projects.

V. Course Content or Outline:
Introduction to the course (one class period)
Rare books and manuscripts (three class periods)
Preservation and conservation (five class periods)
Traditions and principles of archives management (three class periods)
Archives and society (four class periods)
Appraisal and acquisition of archives and manuscripts (one class period)
Processing of archives and manuscripts (three class periods)
Hands-on experience and project work (twelve class periods)

VI. Textbook(s) and/or Other Required Materials or Equipment:
Helene Hanff, *84 Charing Cross Road;* Daniels, ed., *Modern Archives Reader*

VII. Basis for Student Evaluation:
Students will be evaluated on their performance on one mid-term examination and one final examination, one written report, one oral report, contribution to a class project, participation in classroom discussion, hands-on activities and field trips.

Grades will be determined on the following basis:

Mid-term 60	270+ = A
Oral report 30	240+ = B
Written report 50	210+ = C
Project work & report 70	180+ = D
Class participation 30	
Final 60	
TOTAL 300	

Tentative Class Schedule

The Love of Books
Jan. 22 -Feb. 3 Readings, lecture, discussion, and written report on rare books and manuscripts.

Preservation and Conservation
Feb. 5-Feb. 21 Lecture, discussion, and hands-on practice in preservation and conservation.

Modern Archives
Feb. 24- April 4 Lecture, readings, and oral reports on archival concepts, principles, and traditions. Assignment: Each student is to select a country and prepare an oral report on its traditions of archives and record-keeping. The Internet will be a useful source of information. But students should also use printed sources.

Readings from *A Modern Archives Reader* will be assigned.

Mid-term examination: March 14
Spring break, Mar. 17-21.
Oral reports will be scheduled between March 24 and April 4.

Archives and Public Service
Lectures, readings, project work, review.

Final examination, 2 p.m. Monday May 12.

Current Syllabi, 2020-2021

A few samples of the latest, or most current HP curricula, courtesy Dr. Hoffman and other HP professors, follows. Not only will this serve to document a benchmark of scholarship taking place "now," but it also will memorialize how COVID-19 impacted instruction for students and faculty having to shift on a dime to accommodate a global pandemic with an uncertain lifespan. The following syllabi are in order by course number.

Course Syllabus
HP100: Introduction to Historic Preservation
Southeast Missouri State University, Fall 2021
M, W, F 10:00-10:50, CR109

Dr. Steven J. Hoffman
Department of History
311 T Carnahan Hall
Office: (573) 651-2808
Cell: (573) 225-8393

Office Hours: M: 12:00-1:00; W: 11:00-1:00
and by appointment
email: shoffman@semo.edu

www.facebook.com/hoffman.steven

Graduate Assistant:
Zane Peterman
Carnahan 311M
Cell: (573) 201-8046

Office Hours: T 11:00-12:00; Th 11:00-1:00
and by appointment
email:zkpeterman@semo.edu

Course Description

This course is a survey of historic preservation in the United States and examines the theory, philosophy, and methods of maintaining the culture of the past. (3 credit hours)

Purposes and Objectives of the Course

A. To introduce students to the history and theory of historic preservation.
B. To familiarize students with the broad spectrum of activities comprising the work of historic preservationists in particular and public historians more generally.
C. To help students develop skills in observing the built environment.
D. To enable students to develop and improve historical research and communication skills.
E. To provide students with the opportunity to reflect on professional goals and opportunities.

Course Learning Outcomes

A. Students will be able to describe the development of the preservation movement in the United States from its beginnings in the mid-19th century to passage of the National Historic Preservation Act of 1966.
B. Students will be able to demonstrate a familiarity with several basic architectural features.
C. Students will be able to identify at least three (3) alternate careers for historians outside the traditional career path of teaching and/or academic employment and discuss the pros and cons of each.

Expectations of Students

Students are expected to read all assignments, regularly attend class, participate in class discussions and/or activities—including field trips, complete museum/historic site review, a public history project/website, a building survey, an architectural detail scavenger hunt, a short paper, make several oral presentations and successfully complete all exams and written assignments. participation and discussion of the readings form an important part of the course. Students are expected to actively participate in class discussion, including commenting on ideas from the reading assignments, concepts presented by other students, and ideas advanced by the instructor.

Basis for Student Evaluation

Students will be evaluated on class participation--including volunteering outside of class, three projects, five précis, one short paper, a midterm examination, a final examination, and an oral presentation.

Documenting "Place": Developing a Public History collection and outreach blog page (Oral History Project)

Working in groups, students will participate in a public history project to document some aspect of the history of Cape Girardeau. Students will conduct and transcribe video-taped oral interviews of individuals and the resulting interviews will be placed in their larger context, forming the basis for developing a local history blog post or web site. You will be asked to write a short reflective essay about this experience.

Historic Buildings Survey

Students will form teams of two and complete building surveys of both a residential and commercial structure in Cape Girardeau. Surveys should be approximately 1 or 2 paragraphs in length for each structure and provide a brief physical description that highlights the character-defining features of the buildings. The survey will include both a paper report on both buildings and a brief PowerPoint presentation of approximately 5 minutes. Written reports will be due on the day of your presentation.

Panel Discussion Precis

Students will write several 1-page précis of selected articles, which will form the foundation for panel discussions during class. Students should be sure to identify the thesis of the article, as well as describe the most relevant supporting examples. Students should also include their own analysis of the article.

Architectural details scavenger hunt

In order to familiarize themselves with architectural details and terminology, students will complete an architectural detail scavenger hunt.

Museum/Historic Site Review

Students will conduct an analysis of either a museum, museum exhibit, or historic site and prepare a 2-3 page written review.

Volunteer Work

Students will volunteer with a local history or historic preservation organization or group for a minimum of 20 hours. All volunteer work must be approved in advance by the instructor. To receive credit students must turn in a signed statement from their volunteer supervisor.
Note: 5 hours of volunteer time must be submitted no later than the midterm exam due date.

Protect the Nest Classroom Guidelines [See syllabi below for more on these areas]
Note on Plagiarism
Course Policies and Procedures
Attendance
Academic Honesty
Accessibility
Southeast Missouri State University and the Office of Accessibility Services are
Civility
Mandatory Reporting
Student Success
Technology
Honors Program
University Policy Statement
Cell Phone/Computer/Tablet Use in Class

The components of your grade will be weighted as follows:

Midterm Exam: 15%
Final Exam: 15%
Building Description/Presentation 10%
Architectural Detail Scavenger Hunt10%
Public History Project/Blog10%
Panel Discussion Precis 10%
Volunteer Work 10%
Museum/Historic Site Review 10%
Participation, Quizzes, Homework, in-class assignments: 10%

Textbooks(available from Textbook Services and/or Southeast Bookstore)

Gardner, James B. and LaPaglia, Peter S., eds. *Public History: Essays from the Field*. Malabar, FL: Krieger Publishing Company, revised edition, 2004.

Murtagh, William T. *Keeping Time: The History and Theory of Preservation in America* (Third Edition).New York: John Wiley and Sons, Inc., 2006.

Page, Max and Randall Mason, *Giving Preservation a History: Histories of Historic Preservation in the United States*. 2nd Edition. New York: Routledge, 2020.

NOTE: Additional readings will be made available for reading on the internet or photocopied and handed out in class.

Course Outline and Schedule

Date	Day	Topic
Aug. 23	M	Introduction
Aug. 25	W	Part I: Introduction: Historic Preservation and Public History What is Historic Preservation? Assignment: Murtagh, Preface(s) and Introduction.

Aug. 27	F	Historic Preservation Program Standards (NCPE) Assignment: TBA
Aug. 30	M	Overview of Public History Assignment: Patricia Mooney-Melvin, "Professional Historians and the Challenge of Redefinition," pp. 5-22 in Gardner & LaPaglia, *Public History: Essays from the Field;* and Constance B. Schulz, "Becoming a Public Historian," pp. 23-40 in Gardner & LaPaglia, *Public History: Essays from the Field.*
Sept. 1	W	The Language of Preservation Assignment: Murtagh, chapter 1.
Sept. 3	F	Part II: Brief History of Historic Preservation Movement Preservation's Early History Assignment: Murtagh, chapter 2.
Sept. 6	M	Labor Day --No Class
Sept. 8	W	The National Trust and the Federal Involvement in Preservation Assignment: Murtagh, chapters 3 and4.
Sept. 10	F	Preservation Since World War II Assignment: Murtagh, chapter 5.
Sept. 13	M	Part III: Understanding the Built Environment Historic Districts Assignment: Murtagh, chapter 8.
Sept. 15	W	Neighborhood Walking Tour Assignment: TBA
Sept. 17	F	Oral History Project Assignment: TBA
Sept. 20	M	Describing a Building Assignment: Pages 28-34 from National Register Bulletin 16a.(available online either from course website or from National Park Service website).
Sept. 22	W	Preservation in Action—Cape Girardeau Common Pleas Courthouse City Hall project Joy Coleman, Preservation Architect, Treanor HL architects Optional afternoon tour of city hall project
Sept. 24	F	 Dating Buildings Assignment: TBA
Sept. 27	M	Building presentations
Sept. 29	W	Building presentations
Oct. 1	F	Building presentations
Oct. 4	M	Preservation in Action—Internship Reports Rachel Kincy, Hunter Dawson State Historic Site
Oct. 6	W	Preservation in Action –Field Trip Broadway and Esquire Theaters
Oct. 8	F	No Class—Fall break
Oct. 11	M	Parts V: Alternative Perspectives on the History of the Preservation Movement/Careers for historians in Public History Rethinking the Roots of the Historic Preservation Movement Assignment: Page and Mason, *Giving Preservation a History,*

		Introduction," pp. 3-16. DUE: Midterm Exam
Oct. 13	W	Museum Funshop1: Artifacts and Reproductions in a Museum Setting Note: Meet in Museum History gallery at Crisp Museum, River Campus
Oct. 15	F	Preservation in Action –Field TripAcademic Hall
Oct. 18	M	Preservation History Panel Discussion: North and South Assignment: Holleran, Chapter 2in Page and Mason, *Giving Preservation a History,* pp. 17-42; and Lindgren, Chapter 3 in Page and Mason, *Giving Preservation a History*, pp. 43-65. DUE:Precis of your article (either chapter 2OR chapter3, not both)
Oct. 20	W	Online assignment (Jeremy Wells/Bill White: structural racism and CRM) Class does not meet today.
Oct. 22	F	Preservation in Action –National Park Service Guest Speaker: Chris Collins, Superintendent, Ste. Genevieve NHP
Oct. 25	M	Archives I Roy H. Tyron, "Archivists and Records Managers," pp. 57-74 in Gardner &LaPaglia, Public History: Essays from the Field; and Debra Newman Ham, "Manuscript Curators and Specialists," pp. 169-186 in Gardner & LaPaglia, Public History: Essays from the Field. Note: Meet in Kent Library,KL215
Oct. 27	W	Archives II Note: Meet in Kent Library, 3rdFloor, KL306, Special Collections and Archives
Oct. 29	F	Preservation History Panel Discussion: New York and Rhode Island Assignment: Mason, Chapter 4in Page and Mason, *Giving Preservation a History*, pp. 67-99; and Greenfield, Chapter 5 in Page and Mason, *Giving Preservation a History*, pp. 101-122. DUE: Precis
Nov. 1	M	Museums Assignment: Robert B. Patterson, Jr. "In Local Historical Agencies, Museums, and Societies," pp. 295-306 in Gardner & LaPaglia, *Public History: Essays from the Field.* Guest Speaker: Carla Jordan, Director, Cape Girardeau County History Center, Lutheran Heritage Center and Museum, and famous Southeast HP graduate.

Nov. 3	W	Museum Funshop2: Models and Mounts Note: Meet in Museum History gallery at Crisp Museum, River Campus
Nov. 5	F	House and Outdoor Museums/Historic Site Administration Assignment: Murtagh, chaps 6 and 7. Guest Speaker: Donna Rausch, Supervisor and Site Director, Felix Valle Home State Historic Site, and famous Southeast HP graduate.
Nov. 8	M	Online assignment(Brent Leggs, Executive Director, African American Cultural Heritage Action Fund) Class does not meet today.
Nov. 10	W	Preservation History Panel Discussion: Charleston and LGBTQ Assignment: Yuhl, Chapter 8in Page and Mason, *Giving Preservation a History*, pp. 199-225; and, Dubrow, Chapter 12 in Page and Mason, *Giving Preservation a History*, pp. 327-367. DUE: Precis
Nov. 12	F	HP Alumni Networking Workshop
Nov. 15	M	Online assignment (African American and Women's Archives) Class does not meet today.
Nov. 17	W	Museum Funshop3: More Models and Mounts Note: Meet in Museum History gallery at Crisp Museum, River Campus
Nov. 19	F	Public History Panel Discussion Michael Devine, "Administrators: Students of History and Practitioners of the Art of Management," pp. 45-56 in Gardner & LaPaglia, *Public History: Essays from the Field*; Sylvia K. Kraemer, "Policy Advisors: Historians and Making Policy," pp. 217-228 in Gardner& LaPaglia, *Public History: Essays from the Field*; Elizabeth Adkins, "In Business and Corporations: Serving as the Corporate Memory," pp. 371-384 in Gardner & LaPaglia, *Public History: Essays from the Field*; and Philip L. Cantelon, "As a Business: , Not Bought," pp. 385-395 in Gardner & LaPaglia, *Public History: Essays from the Field.* DUE:Precis
Nov. 22	M	Choose your own adventure: Public History career options Assignment: Choose one of the chapters in Gardner & LaPaglia, Public History: Essays from the Field not previously assigned this semester, write a précis on it PLUS adda paragraph on why you chose the article, and how it informs your career plans. Post to Moodle forum. In addition to your original post, you must comment on at least one a classmate. Class does not meet today. DUE: Precis DUE:Architectural detail scavenger hunt

Nov. 24	W	Thanksgiving Holiday —No Class
Nov. 26	F	Thanksgiving Holiday —No Class
Nov. 29	M	Part VI: Built Environment Preservation Revisited Rehabilitation and Adaptive Use Assignment:Murtagh, chapter 9.
Dec. 1	W	Main Street and Economic Redevelopment Assignment: The Main Street Approach: A Comprehensive Guide to Community Transformation, pp. 1-14. Guest Speaker: Sarah LaVenture, Assistant Director, Old Town Cape, Cape Girardeau Main Street Program, and famous Southeast HP graduate.
Dec. 3	F	Preservation in Action—Internship Reports Andrea Hamm
Dec. 6	M	Preservation Today and Tomorrow Assignment: Kaufman, Conclusion in Page and Mason, *Giving Preservation a History*, pp. 391-407. DUE: Museum Exhibit/Historic Site Review
Dec. 8	W	Where to from Here? Assignment: Read: Page, “Epilogue: Not Your Grandmother’s Preservation Movement,” pp. 409-418.Conduct Internet Research on most important issues in Preservation DUE: 1-2 page paper on 3 most important issues in Historic Preservation
Dec. 10	F	Becoming a Professional: Reflectionsand Conclusions DUE: Oral HistoryProject DUE: Oral History Project Reflective Essay
Dec. 13	M	HP100 final Take Home Final Exam Final Exam due in Canvas by 12 pm

Instructor: Dr. Lily Santoro	Teaching Assistant: Augusta Welsh
Email: lsantoro@semo.edu	Email: ajohnson12s@semo.edu
Office: Carnahan 311C	Office: Carnahan 311M
Hours: M 1:00-3:00; T 2:00-5:00	Hours: T/Th 2:00-3:30
Students can also meet with Dr. Santoro or Ms. Welsh by appointment in person or on Zoom.	

Introduction to Techniques of Local History (HP200)

Spring 2021 ~ Tues. & Thurs. 12:30-1:45 ~ Carnahan 109

CATALOG DESCRIPTION:

Introduction to value, techniques and methods of local history for the preservationist (3 credit hours). Prerequisite: HP100 or consent of instructor.

MEETING LOCATION NOTE:

This is a Mixed Format course: Due to social distancing classroom limitations, students will be broken into two attendance groups (Group 1 and Group 2). Students are expected to attend in class, on Zoom, or at another location in accordance with their attendance group assignments (as indicated in the course schedule).

Attendance group assignments are posted on the course page in Canvas: https://semo.instructure.com

COURSE OBJECTIVES:

1. Provide students with an understanding of local history and its relation to national issues and themes.
2. Introduce students to sources and techniques that are necessary for developing materials used in archives, museums, historic sites, and in restoration.

Class meetings consist of occasional brief lectures and seminar discussions of assigned readings. Writing and research exercises are emphasized, and there will be several projects, field trips, and guest lectures throughout the course. Each student will complete a local history research project to be presented in written and visual formats at the end of the semester.

COURSE LEARNING OUTCOMES:

1. Demonstrate knowledge of major problems and issues in the practice of local history.
2. Locate, gather and assess relevant primary and secondary sources, including archival sources.
3. Analyze and interpret primary and secondary sources using established procedures and criteria in the discipline of history.

REQUIRED MATERIALS:

There are two required textbooks, available for purchase at the University Bookstore.

Howe, Barbara and Dolores A. Fleming, Emory L. Kemp, Ruth Ann Overbeck. *Houses and Homes: Exploring their History*. Lanham, MD: Rowman & Littlefield/AASLH, 1997.

Listed in Course Schedule as **"HOUSES."**

Bacon, Reginald W. *The Micro-historians Guide to Research, Evidence, and Conclusions.* Lanham, MD: Rowman & Littlefield, 2019.

Listed in Course Schedule as **"MICRO."**

Additional readings will be available on Canvas and/or through Kent Library Ebooks and Databases. Students can find these resources at:

https://semo.instructure.com

http://library.semo.edu/

Students are expected to keep up with the assigned readings in the textbooks as well as complete any other reading assignments listed in the syllabus or given in class. Readings are listed for each class meeting in the Course Schedule/Assignments section of this syllabus. **Students are expected to have read the assignment for that day *before* attending class.** Students *must* bring readings to class on the day they are assigned.

COURSE ASSIGNMENTS:

Detailed instructions for all assignments in this course will be available on Canvas. A brief description of each grade component is as follows:

PARTICIPATION: Students are expected to attend class and participate in class discussion of readings and lecture content (online and in person). ***Not attending or participating in class will hurt your grade***. So, avoid absences and be prepared to discuss the readings and other class assignments. All in-class exercises will count toward this portion of your grade. Participation grades are dependent upon actual participation (silent attendance does not earn credit; infrequent attendance/participation will be penalized). Participation will count for 10% of your course grade.

CITING YOUR SOURCES LESSON: Because this is a writing-intensive course, all students are expected to adhere to the strictest guidelines regarding academic honesty. All students must complete an online lesson on plagiarism and citations. The lesson is available in Canvas. The lesson is untimed, and students may complete it as many times as they like but must pass it by 11:59pm on Feb. 5 to receive credit. All students must earn 100% to earn credit for this lesson. The lesson is worth 3% of your course grade.

BOOK REVIEWS: Students must write two book reviews (two-pages each) on secondary sources from the bibliographies in either of the assigned textbooks. Detailed guidelines will be available on Canvas. Each book review will be worth 6% of your course grade (12% total).

MIDTERM EXAM: There will be only one Midterm exam in this course. There is no Final Exam. The Midterm Exam will be essay format and should be written in a blue book. The Midterm Exam will be worth 15% of your course grade.

RESEARCH PROCESS ASSIGNMENTS: Throughout the semester, students will be required to demonstrate their progress on their local history projects by submitting required pieces of their research process. There are four research process assignments: Mobius/ILL Printout, Annotated Preliminary Bibliography, Preliminary Research Notes, and Newspaper Search. Instructions for each assignment will be posted in Canvas. Due dates for each assignment are listed in the Course Schedule below. Each of these assignments will be worth 5% of your course grade (20% total).

LOCAL HISTORY POSTER: Students will present their completed local history research in a public poster session to be held in the University Center on May 4. This is a professional event, for which students should prepare a visual representation (poster) of their research to share with the public (and the rest of the class) and a short speech (approx. 5 minutes) explaining their project. Additional instructions and details for the poster session will be available on Canvas and discussed in class. Posters and presentations are worth 20% of your course grade.

FINAL PAPER: In addition to presenting individual projects visually and orally in the poster session, each student must submit a final paper. This is to be a complete research paper, giving the history of the topic with relevant research, images, analysis, and accurate citations. Complete instructions for the final paper will be available on Canvas and will be discussed in class. The final paper is worth 20% of your course grade.

HOW GRADES ARE DETERMINED:

Participation	10%
Citing Your Sources Lesson	3%
Book Reviews	12%
Midterm Exam	15%
Research Process Assignments	20%
Poster	20%
Final Paper	20%
Total	100%

ASSIGNMENT GRADE SCALE:

A=93-100; A-=90-92.99; B+=87-89.99; B=83-86.99; B-=80-82.99; C+=77-79.99; C=73-76.99; C-=70-72.99; D+=67-69.99; D=63-66.99; D-=60-62.99; F=0-59.99

COURSE GRADE SCALE:

A=90-100; B=80-89; C=70-79; D=60-69; F=0-59

LEARNING ENVIRONMENT EXPECTATIONS:

CLASSROOM GUIDELINES FOR MINIMIZING THE RISK OF COVID-19:

- Practice social distancing (six feet of spacing), wear facial coverings, and follow proper prevention hygiene, such as washing your hands frequently and using alcohol-based (at least 60% alcohol) hand sanitizer when soap and water are not available.
- *The wearing of a face covering is a University safety requirement under the current conditions as well as a Cape Girardeau County Emergency Order pursuant to RSMo 192.300. If a student refuses to wear a face covering, it would be handled in the same manner as if they refused to comply with a course requirement or adhere to the Code of Student Conduct. An instructor may ask a student to leave a class and an employee may ask a person to leave a meeting or event if they do not wish to comply with the Face Coverings and Social Distance Guidelines.*
    - *In the event that a student does not comply with this guidance and does not leave the space, faculty/staff should send the student's name and email address or Southeast ID to deanofstudents@semo.edu. If a student refuses to leave a classroom or other University facility upon request,* ***and/or becomes disruptive,*** *the Department of Public Safety can be contacted for assistance.*

3

- **If you are not able to wear face coverings due to health or ADA concerns, contact the office of Disability Services at ds@semo.edu or 573-651-5927 for information on how to obtain an exemption. Affected individuals should also consult with their physicians before wearing a face covering.**

- Avoid sharing electronic devices, books, pens, and other learning aids unless thoroughly cleaned between users.
- Students are encouraged to clean and disinfect, before and after class, any contact surfaces such as chairs, desks, and equipment utilized for the class. Students will follow the protocol established for this classroom.
- Appropriate signage will be displayed in the classroom or adjacent areas on protocols to follow that protect our community members during classes. Please become familiar with this information and follow it in our shared community.
- The seating in your classroom may be positioned to maintain social distancing or signs may be posted with seating direction. Do not move seating or signs unless instructed to do so by your instructor.
- Social Distancing minimizes risk. When in buildings, hallways and entering or exiting buildings or classrooms, maintain at least six feet of social distancing space and avoid gathering in groups.
- To minimize risk and to follow the face covering guidelines, there should be no eating in the classrooms or indoor instructional spaces. Drinks are permissible provided face coverings are removed only while drinking and are properly worn when not drinking. If other guidelines regarding no food or drink are posted or conveyed, those guidelines must be followed. Medical necessities for food should be communicated to the instructor and are permissible. Students should minimize the time without a face covering as much as possible.

CLASSROOM DISCUSSION & CIVILITY: All students are expected to be active participants in class, and join in polite discussion (no name-calling, slurs, or talking over each other). Because historical actors often have funny names and unusual worldviews, it is understandable if you mispronounce names, places, and events. There is even room for lighthearted discussion of historical actors' motives and actions. However, students must exhibit respect for their classmates and instructors (and their opinions/questions/answers) at all times.
Remember, you are responsible for your behavior and actions. There is a no tolerance policy on bullying or harassment of any kind. Additional information on student conduct may be found at: http://www.semo.edu/pdf/stuconduct-code-conduct.pdf?ver=1.0 and http://www.semo.edu/pdf/Conduct_Faculty_Resource_Guide.pdf

CELL PHONE POLICY: While accidents happen (cell phones ring, etc.), **text messaging, video games, facebook, "google-ing," etc. are NOT allowed during class**. Unless specifically instructed to do so, **students may not use their phones during class**. If you are using your phone during class, you will be counted as absent for the day (negating any participation points earned that day). **Cell phones and all other digital devices must be turned *OFF* during the midterm exam** (not silent, not vibrate, not airplane mode). If you cannot turn your device off, you should not bring it to the test. If your device rings or receives (or sends) a text message during an exam, your exam will be taken away and graded as-is. NO EXCEPTIONS.

ACADEMIC HONESTY: Southeast Missouri State University expects all students, faculty and staff to operate in an honest and ethical manner. Academic dishonesty is a very serious offense because it undermines the value of your education and the education of others. Students who engage in academic dishonesty face significant penalties. Forms of academic dishonesty include, but are not

4

limited to, plagiarism, cheating, contract cheating, misrepresentation, and other actions you take. Some of these are defined below:

- Plagiarism means passing off someone else's work as your own, whether it is intentional or unintentional.
- Cheating includes copying from another person or source of information to meet the requirements of a task.
- Contract cheating is paying someone else or a company to do your work.
- Misrepresentation means you are posing as someone else or someone else is posing as you to complete a task.
- Collusion means working with one or more people to cheat. If you help someone cheat or plagiarize you will face the same penalties.

For more information, visit the Responsible Redhawks Code of Conduct http://www.semo.edu/responsibleredhawks/code-of-conduct.html or the Faculty Handbook Section (D) on Academic Honesty http://www.semo.edu/facultysenate/handbook/5d.html

All students who violate the University's academic honesty policies in this class will receive a ZERO for the assignment (without the opportunity to make it up) and may be reported to Student Conduct. Make sure the language and analysis you use in your written assignments is your own. If you are using someone else's ideas or insights, provide accurate citations and quotation marks. If you are unsure, consult with the professor ***before*** turning in the assignment. It is always best to err on the side of caution when dealing with questions of academic honesty. **Plagiarism and academic dishonesty will not be tolerated in this course. No exceptions will be granted.**

LATE WORK/MISSED EXAMS: In *extremely unusual* circumstances (excused absences with written verification), students may need to make up a missed exam. Should this happen, the student is responsible for contacting the professor and making arrangements for a make-up exam as soon as possible.

In-class exercises and poster presentations cannot be made up.

Late papers will not be accepted.

THE WRITING LAB: The Writing Lab is located in Memorial Hall 103, where it provides face-to-face and online writing support. For hours, see: http://www.semo.edu/writing/writing_lab.htm . The tutors in the Writing Lab will help you with your writing at any stage of the process, both on a drop-in and appointment basis. What the Writing Lab tutors will NOT do is simply proofread your paper. The Online Writing Lab (http://ustudies.semo.edu/writing/owl.asp) is also available to provide similar services.

Phone: (573) 651- 2460 | Location: Memorial Hall 103

TECHNOLOGY: All written assignments must be typed, using a word-processing program, such as MS Word. Students will have the opportunity, but will not be required, to submit some written assignments electronically in Canvas. All other assignments must be turned in as hard-copies (printed on paper) at the beginning of class on the due date. This means students must have access to a printer. Students will be required to access some of the reading material for the course on the internet. Students who anticipate difficulty in mastering any of the technological components of the course are responsible for seeking assistance from the instructor in a timely manner.

The primary mode of extra-classroom communication for this course will be through University-provided email. All students are expected to check their @semo.edu email accounts for updates and messages regularly (typically once a day).

COMMUNICATION EXPECTATIONS:

Questions, comments or requests regarding this course or program should be taken to your instructor, Dr. Lily Santoro. Unanswered questions or unresolved issues involving this class may be taken to Dr. Toni Alexander, Chair, Department of History & Anthropology, who can be contacted at 573-651-2478.

OFFICE HOURS: Dr. Santoro and Ms. Welsh will hold drop-in office hours in Carnahan Hall on during the hours listed at the top of this syllabus.

If you prefer to talk by phone, you may call Dr. Santoro *during office hours*: (573) 651-2831.

Students who prefer a video chat may meet with the instructor on Zoom *by appointment*. Appointments can be arranged by sending an email to Dr Santoro (lsantoro@semo.edu) or Ms. Welsh (ajohnson12s@semo.edu).

DIRECT COMMUNICATION WITH INSTRUCTORS: The primary mode of personal communication with the instructors for this course is email. Typically, we will respond to emails within 24 hours, Mon-Fri, and within 48 hours on weekends and during University holidays. Because class related communication with students is only authorized via official SEMO email addresses, students must use their @semo.edu email to contact Dr. Santoro or Ms. Welsh and should check that email daily.

CLASS ANNOUNCEMENTS: Class-wide announcements will be posted in the Announcements on our Canvas page. Most students receive these announcements in their @semo.edu email, but you should check that forum regularly to be sure you are caught up.

CONFIDENTIALITY AND MANDATORY REPORTING: Instructors will keep information you share with me confidential to the best of my ability, but as a professor/TA we are legally required to share information about sexual misconduct and crimes we learn about to make our campus and community safe for everyone.

STUDENTS WITH DISABILITIES:

Southeast Missouri State University and Disability Services are committed to making every reasonable educational accommodation for students who identify as people with disabilities. Many services and accommodations which aid a student's educational experience are available for students with various disabilities. Students are responsible for contacting Disability Services to register and access accommodations. Accommodations are implemented on a case by case basis. For more information, visit http://www.semo.edu/ds/ or contact Disability Services at 573-651-5927.

COURSE SCHEDULE:

Course schedule may be subject to change. Any changes will be announced in class and Canvas.

Link for Zoom attendance: https://semo.zoom.us/j/99409954783

Week 1 What is Local History?

Jan 26 **Group 1: CR019** **Group 2: ZOOM**

Course Introduction & Activity

Jan 28 **Group 1: ZOOM** **Group 2: CR109**

What is Local History?

Read:

MICRO, pp. 93-98 (Appendix I: Local History 101)

Week 2 Setting the Parameters of Local History

Feb 2 **Everyone: ONLINE FORUM DAY** (no in-person class)

Find an online history of your hometown or county and post a link to that resource in Canvas by 12:30pm on Tuesday. Please review at least two of your classmates' websites in the forum by 5pm on Wednesday.

Feb 4 **Group 1: ZOOM** **Group 2: CR109**

Whom are we studying? How do we know the past?

Read:

MICRO, pp. 11-26; pp. 49-72 (Chapter 1: Research Planning; Chapter 4: Formulation of Conclusions)

HOUSES, pp. 1-6 (Chapter 1)

*****Remember to complete the Citing Yours Sources Lesson by 11:59pm on Friday!**

Week 3 Beginning Research: Theory and Practice

Feb 9 **Group 1: CR109**

Getting Started at Special Collections & Archives

Guest Speakers:

Roxanne Dunn & Tyson Koenig
Special Collections and Archives

Students are encouraged to bring laptops or tablets for today's in-class activity. Tablets will be provided in class for students who do not have a device to use.

Group 2: Kent Library 215

Starting Your Research at Kent Library

Guest Speaker:

Karl Suhr, Information Services/Information Literacy

7

Feb 11 **Group 1: Kent Library 215**

Guest Speaker:

Karl Suhr, Information Services/Information Literacy

Group 2: CR109

Getting Started at Special Collections & Archives

Guest Speakers:

Roxanne Dunn & Tyson Koenig
Special Collections and Archives

Students are encouraged to bring laptops or tablets for today's in-class activity. Tablets will be provided in class for students who do not have a device to use.

*****Remember to sign up for a paper topic by 11:59pm on Monday, Feb. 15!**

Week 4 Written Sources

Feb 16 **Group 1: CR109**

Finding, Evaluating, and Using Written Records

Read:

HOUSES, pp. 39-58 (Chapter 4)

MICRO, pp. 27-48 (Chapter 2: Research Execution & Chapter 4: Evaluation of Evidence)

Group 2: CAPE CO. ARCHIVES CTR

Using the County Archives

Guest Speakers:

Marybeth Neiderkorn & Lyle Johnston

Cape Girardeau County Archive Ctr
112 East Washington
Jackson, Missouri 63755

Feb 18 **Group 1: CAPE CO. ARCHIVES CTR**

Using the County Archives

Guest Speakers:

Marybeth Neiderkorn & Lyle Johnston

Cape Girardeau County Archive Ctr
112 East Washington
Jackson, Missouri 63755

Group 2: CR109

Finding, Evaluating, and Using Written Records

Read:

HOUSES, pp. 39-58 (Chapter 4)

MICRO, pp. 27-48 (Chapter 2: Research Execution & Chapter 4: Evaluation of Evidence)

Week 5 Visual Sources

Feb 23 **NO CLASS: READING DAY**

Feb 25 **Group 1: CR109**

Evaluating Visual Sources

Read:

HOUSES, pp 9-22 (Chapter 2) & pp. 65-75 ("Visual Documents")

Group 2: ZOOM

Catherine Armstrong, *Using Non-Textual Sources: A Historian's Guide* Chapter 2: "Reading Images," pp. 41-72. ***(E-Book available via Kent Library)***

"Cape Girardeau History and Photos" website available at: http://www.capecentralhigh.com

******Research Process Assignment 1: Mobius/Inter-Library Loan Screenshot due***

Week 6 Oral Sources

Mar 2

Group 1: CR109

Using Oral Sources

Read:

HOUSES, pp. 59-65 (Chapter 5 – "Oral Sources")

Kate Fisher, "'She was quite satisfied with the arrangements I made: Gender and Birth Control in Britain 1920-1950," *Past & Present*, No. 169 (Nov., 2000), pp. 161-193.
(available on JSTOR)

Group 2: Kent Library 215

Keeping All This Stuff Straight: Tips on Notetaking

Guest Speaker:

Karl Suhr, Information Services/Information Literacy

Mar 4

Group 1: Kent Library 215

Keeping All This Stuff Straight: Tips on Notetaking

Guest Speaker:

Karl Suhr, Information Services/Information Literacy

*****Book Review #1 due***

Group 2: CR109

Using Oral Sources

Read:

HOUSES, pp. 59-65 (Chapter 5 – "Oral Sources")

Kate Fisher, "'She was quite satisfied with the arrangements I made: Gender and Birth Control in Britain 1920-1950," *Past & Present*, No. 169 (Nov., 2000), pp. 161-193.
(available on JSTOR)

Week 7 Homes: Components, Types & Techniques

Mar 9

EVERYONE: Dempster 127

The American Dream in Housing

Read:

HOUSES, pp. 77-106 (Chapters 6 & 7)

Guest speaker:

Dr. Steven Hoffman, Historic Preservation Program Coordinator

Mar 11	**Group 1: ZOOM** *Optional Midterm Q&A Session*	**[Group 2: CR109** Houses and Building Materials Read: HOUSES, pp. 107-124 (Chapter 8) Guest speaker: Dr. Steven Hoffman, Historic Preservation Program Coordinator

*****Research Process Assignment 2: Annotated Preliminary Bibliography due***

Week 8 Midterm Week

Mar 16	**Group 1: CR109** Houses and Building Materials Read: HOUSES, pp. 107-124 (Chapter 8) Guest speaker: Dr. Steven Hoffman, Historic Preservation Program Coordinator	**Group 2: ZOOM** *Optional Midterm Q&A Session*
Mar 18	**EVERYONE: Dempster 127** MIDTERM EXAM *** ***Research Process Assignment 3: Preliminary Research Notes due***	

Week 9 Poster and Citation Nuts and Bolts

Mar 23	**Group 1: CR109** Bibliographies and Footnotes Please become familiar with the Chicago Manual of Style. Students may use smartphones, laptops, or tablets for today's in-class activity. Tablets will be provided in class for students who do not have a device to use.	**Group 2: Kent Library TBA** Designing Posters for Public and Academic History Guest Speaker: Roxanne Dunn, Special Collections & Archives
Mar 25	**Group 1: Kent Library TBA** Designing Posters for Public and Academic History Guest Speaker:	**Group 2: CR109** Bibliographies and Footnotes Please become familiar with the Chicago Manual of Style. Students may use smartphones, laptops, or tablets for

10

	Roxanne Dunn, Special Collections & Archives	today's in-class activity. Tablets will be provided in class for students who do not have a device to use.

Week 10 Have you been researching???

Mar 30 - Apr 1	NO CLASS this week—Work on your projects and meet with instructors. All students must sign up for a meeting with Augusta Welsh or Dr. Santoro this week to discuss their research progress. Signup and available times will be posted in Canvas.

Week 11 Using Local History Sources

Apr 6	**Group 1: CR109** Using Newspapers and County Records	**Group 2: Kent Library TBA** Using Historical Maps at Special Collections & Archives
Apr 8	**Group 1: Kent Library TBA** Using Historical Maps at Special Collections & Archives ******Book Review #2 due***	**Group 2: CR109** Using Newspapers and County Records

Week 12 Researching, Writing and Presenting History Locally

Apr 13	**Group 1: CR109** Cultural Resource Management and Local History for Academia Read: MICRO, pp. 73-90 (Chapter 5: The Summary Narrative) Additional online readings available in Canvas	**Group 2: FIELD TRIP** Visit to the Cape River Heritage Museum 538 Independence St. Cape Girardeau, MO 63701
Apr 15	**Group 1: FIELD TRIP** Visit to the Cape River Heritage Museum 538 Independence St. Cape Girardeau, MO 63701	**Group 2: CR109** Cultural Resource Management and Local History for Academia Read: MICRO, pp. 73-90 (Chapter 5: The Summary Narrative) Additional online readings available in Canvas

******Research Process Assignment 4: Newspaper Search Assignment due***

Week 13 Are you writing????

Apr 20-22 NO CLASS—finish up your posters and projects

******Digital Poster files must be uploaded to Canvas by 5pm on Thurs., April 22.***

Week 14 Presenting History Locally

Apr 27

Group 1: CR109

Writing, presenting and evaluating "good" local history

Read:

HOUSES, pp. 147-158 (Chapter 11)

Group 2: FIELD TRIP

Visit to the Red House Interpretive Center

211 Water St.
Cape Girardeau, MO 63701

Apr 29

Group 1: FIELD TRIP

Visit to the Red House Interpretive Center

211 Water St.
Cape Girardeau, MO 63701

Group 2: CR109

Writing, presenting and evaluating "good" local history

Read:

HOUSES, pp. 147-158 (Chapter 11)

******Last day to get credit for your Mobius/ILL book.***
(Make sure you have shown it to Dr. Santoro or Augusta Welsh by the end of class.)

Week 15 Poster Session Week

May 4 **EVERYONE: UNIVERSITY CENTER (room TBA)**

The class poster session will be held in the University Center during our usual class period.

Please consult the instructions on Canvas for more details.

May 6

Group 1: ZOOM

Group 2: CR109

Wrapping up

Today's class period will be an opportunity to discuss any questions or concerns you have about the final project or other course content. Even if you don't have questions, come ready to share your knowledge with those who do.

Week 16 Finals Week

May 11 Final Paper due by 11:59pm on Tuesday, May 11.

Everyone must submit their final paper in person or to the Dropbox in Canvas by 11:59pm on Tuesday, May 11.

Late Papers will not be accepted. Plan ahead!

Issues in Historic Preservation: Archives and Special Collections Management (HP630)

~~~~~~~~~~~~

Archives and Special Collections Management (HP405)

~~~~~~~~~~~~

Spring 2021 ~ Serena 312 ~ M/W 3:00-4:15

Instructor Information:

Instructor: Dr. Lily Santoro
Email: lsantoro@semo.edu
Office: Carnahan 311C
Hours: Mon. 1:00-3:00 & Tues. 2:00-5:00
Phone: 573.651.2831 (during office hours only)

Course Objectives:

To provide students the opportunity to study topics related to specific areas of public history interests. The specific focus of the course will normally be archives management, museum management, or historic site administration, depending upon the specific expertise of the instructor.

This course will introduce students to the history, theories, and operational issues involved in managing archives and special collections as well as their application to the organization and administration of archival collections within public and private institutions. Through readings, lectures, discussions, and hands-on activities, students will become familiar with the standard practices related to surveys of resources, acquisitions and processing, housing and equipment, conservation and restoration, and user access in archives and special collections.

Course Learning Outcomes for HP630:

1. Students will be able to give examples of, describe, and explain accepted practices and procedure relevant to the preservation, interpretation, presentation, management or administration of historical documents.

2. Students will be able to demonstrate knowledge of relevant issues in the field of archives.

3. Students will be able to communicate, in written and oral forms, using the technical language of the archival professional.

Course Learning Outcomes for HP405:

1. Students will be able to articulate the history and philosophy of care, organization, and conservation of archival collections.

2. Students will be able to demonstrate proper care, conservation, and storage of archival collections.

3. Students will be able to give examples of, describe, and explain accepted practices and procedures of archives and special collections administration.

Required Readings:

There is one required textbook for this course:

Laura A. Millar, *Archives: Principles and Practices.* (2004).

Additional readings listed in the syllabus are available through Kent Library and/or in Canvas. The professor may also give out additional reading assignments in class.

The required books for this course can be purchased at the University Bookstore. There will also be additional readings made available in Canvas. Students are expected to keep up with all assigned readings listed in the syllabus and complete any reading assignments given in class. Readings are listed for each week in the Course Schedule. ***Students are expected to have read the assignment for the week before attending class.***

Basis of Student Assessment:

All students are expected to participate in all class sessions; read all assigned materials; effectively complete the class projects, written and practical, listed below. Those enrolled in HP630 are also required to complete a literature review of a current archival issue. Complete instructions for all assignments are posted in Canvas.

PARTICIPATION & IN CLASS ACTIVITIES:
Students are expected to attend all class sessions and be active participants in all in-class activities and discussion of readings. Students should come to class prepared to take notes and talk about the topics and themes covered in the readings for that week. To do this effectively, you will need to complete the entire reading assignment before attending class for that week. Many reading assignments are long and, occasionally, complex. It is recommended that you take notes while you read and bring any questions to class.
Participation grades are dependent upon actual participation in discussion and activities (silent/uncooperative attendance does not earn credit).
ALL STUDENTS: Participation and in class activities are worth 20% of your final grade.

PAPERS:
Students will write three short papers (1-2 pages) in response to a prompt posted in Canvas. Follow the instructions for each paper closely. Students should submit their papers in Canvas by the beginning of class on the due date. Late papers will not be accepted. Papers that violate the University's academic honesty policy will receive a zero.
HP630: Each paper is worth 5% of your course grade.
HP405: Each paper is worth 10% of your course grade.

CONSERV-O-GRAM PRESENTATION & HANDOUT:
Students will work in pairs to do outside research on a topic covered by the National Parks Service's Conserv-O-Gram and present their findings to the class with accompanying visuals (PowerPoint or similar) and explanatory handout (1 page, double sided) with a bibliography of current scholarship/resources on the topic. All students must sign up for topic in Canvas by Midnight on Friday, February 12 (end of Week 3). Additional instructions for this project will be available in Canvas.
ALL STUDENTS: The presentation and handout will count for 20% of your course grade.

ARCHIVAL ISSUE LITERATURE REVIEW: (HP630 only)
Grad students will choose an archival management issue that is the subject of debate among

2

professionals today (list available in Cnavas). Students will become familiar with the existing scholarship on their chosen issue and write a 5-8 page essay outlining the debates and consensus regarding best practices in the field. In your essay, you should identify the five most important works on this topic and indicate your professional analysis of the issue. Students should submit their papers in Canvas by the beginning of class on the due date. Papers that violate the University's academic honesty policy will receive a zero.
HP630: The literature review is worth 15% of your course grade

FINAL PROJECT:
In the final weeks of the semester, each student will accession and process their own manuscript collection. **Completed projects will consist of a processed collection in proper storage, a descriptive finding aid, and a process paper** (5-6 pages) that summarizes how the student completed the project. Students' papers, finding aids, and processed collections must adhere to the guidelines posted in Canvas and the practices and standards studied in this course.
ALL STUDENTS: The completed final project (collection, finding aid, & paper) is worth 30% of your course grade.

How Grades are Determined:

HP405 (Undergrad)	
Participation & Activities	20%
Paper #1	10%
Paper #2	10%
Paper #3	10%
Conserv-O-Gram	20%
Final Project	30%
Total	100%

HP630 (Grad)	
Participation & Activities	20%
Paper #1	5%
Paper #2	5%
Paper #3	5%
Conserv-O-Gram	20%
Literature Review	15%
Final Project	30%
Total	100%

ASSIGNMENT GRADE SCALE:
A=93-100; A-=90-92.99; B+=87-89.99; B=83-86.99; B-=80-82.99; C+=77-79.99; C=73-76.99; C-=70-72.99; F=0-69.99

COURSE GRADE SCALE:
A=90-100; B=80-89; C=70-79; F=0-69

Learning Environment Expectations:

This course abides by the university's policies regarding attendance, Covid-19 health and safety, academic honesty, civility and harassment, and support for students with disabilities, and has additional requirements regarding the use of technology:

ATTENDANCE & CLASSROOM DISCUSSION:
All students are expected to be active participants in class, and join in polite discussion (no name-calling, slurs, or talking over each other). All students must exhibit respect for their classmates (and their opinions/questions/answers) at all times.

CELL PHONES AND TEXT MESSAGING:
While accidents happen (cell phones ring, etc.), **text messaging, video games, facebook, "google-ing," etc. are NOT allowed during class**. Unless specifically instructed to do so, students may not

3

use their phones during class. If you are using your phone during class, you will be counted as absent for the day (negating any participation points earned that day).

ACADEMIC DISHONESTY:
All students who violate the University's academic honesty policies in this class will receive a zero for the assignment, without the opportunity to make it up, and may be reported to Student Conduct. Make sure the language and ideas you use in your written assignments are your own. If you are using someone else's ideas or insights, provide accurate citations and quotation marks. If you are unsure, consult with the professor before turning in the assignment. It is always best to err on the side of caution when dealing with questions of academic integrity.

Students should be familiar with the University's policy regarding academic honesty:
http://www.semo.edu/facultysenate/handbook/5d.htm

If you are not sure what constitutes plagiarism/academic dishonesty, check out this tutorial provided by the Center for Writing Excellence:
http://ustudies.semo.edu/writing/owl2/tutorials/plagiarism/

CLASSROOM GUIDELINE FOR MINIMIZING THE RISK OF COVID-19

- Practice social distancing (six feet of spacing), wear facial coverings, and follow proper prevention hygiene, such as washing your hands frequently and using alcohol-based (at least 60% alcohol) hand sanitizer when soap and water are not available.
- *The wearing of a face covering is a University safety requirement under the current conditions as well as a Cape Girardeau County Emergency Order pursuant to RSMo 192.300. We hope that through modeling and a positive and encouraging environment, all will do their part. If a student refuses to wear a face covering, it would be handled in the same manner as if they refused to comply with a course requirement or adhere to the Code of Student Conduct. An instructor may ask a student to leave a class and an employee may ask a person to leave a meeting or event if they do not wish to comply with the Face Coverings and Social Distance Guidelines.*

    - *In the event that a student does not comply with this guidance and does not leave the space, faculty/staff should send the student's name and email address or Southeast ID to deanofstudents@semo.edu. If a student refuses to leave a classroom or other University facility upon request,* ***and/or becomes disruptive,*** *the Department of Public Safety can be contacted for assistance.*

    - **If you are not able to wear face coverings due to health or ADA concerns, contact the office of Disability Services at ds@semo.edu or 573-651-5927 for information on how to obtain an exemption. Affected individuals should also consult with their physicians before wearing a face covering.**
- Avoid sharing electronic devices, books, pens, and other learning aids unless thoroughly cleaned between users.
- Students are encouraged to clean and disinfect before and after class, any contact surfaces such as chairs, desks, and equipment utilized for the class. Students will follow the protocol established for this classroom.
- Appropriate signage will be displayed in the classroom or adjacent areas on protocols to follow that protect our community members during classes. Please become familiar with this information and follow it in our shared community.

- The seating in your classroom may be positioned to maintain social distancing or signs may be posted with seating direction. Do not move seating or signs unless instructed to do so by your instructor.
- Social Distancing minimizes risk. When in buildings, hallways and entering or exiting buildings or classrooms, maintain at least six feet of social distancing space and avoid gathering in groups.

ACCESSIBILITY & ACCOMODATIONS:
Southeast Missouri State University and Accessibility Services are committed to making every reasonable educational accommodation for students who identify as people with disabilities. Many services and accommodations which aid a student's educational experience are available for students with various disabilities. Students are responsible for contacting Disability Services to register and access accommodations. Accommodations are implemented on a case-by-case basis. For more information, visit http://www.semo.edu/ds/ or contact Accessibility Services at 573-651-5927.

If you require accommodations for this course, please contact the instructor to make arrangements within the first few weeks of the semester. Accommodations cannot be implemented retroactively.

COMMUNICATION EXPECTATIONS:
Questions, comments or requests regarding this course or program should be taken to your instructor, Dr. Lily Santoro. Unanswered questions or unresolved issues involving this class may be taken to Dr. Toni Alexander, Chair, Department of History & Anthropology, who can be contacted at 573-651-2478.

Office Hours:
Dr. Santoro will hold drop-in office hours in Carnahan 311C on Mondays 1-3 and Tuesdays 2-5. If you prefer to talk by phone, you may also call *during office hours*: (573) 651-2831.
Students who prefer a video chat may meet with the instructor on Zoom *by appointment*.
Appointments can be arranged by sending an email to lsantoro@semo.edu

Direct Communication with Instructor:
Dr. Santoro's primary mode of personal communication is email (lsantoro@semo.edu). Typically, she will respond to emails within 24 hours, Mon- Fri, and within 48 hours on weekends and during University holidays. Because class related communication with students is only authorized via official SEMO email addresses, students must use their @semo.edu email to contact Dr. Santoro and should check that email box daily.

Class Announcements:
Class-wide announcements will be posted in the Announcements on our Canvas page. Most students receive these announcements in their @semo.edu email, but you should check that forum regularly to be sure you are caught up.

Confidentiality and Mandatory Reporting:
I will keep information you share with me confidential to the best of my ability, but as a professor I am legally required to share information about sexual misconduct and crimes I learn about to make our campus and community safe for everyone.

Course Schedule:

Note: Course schedule may be subject to change. Any changes will be announced in class and in Canvas Announcements.

Archives and Special Collections History and Philosophy

Week 1 (Jan 25-27) **Introduction to Archives/Special Collections:**
Read: Chap. 1 "What are Archives?"

Howard Besser. "The Museum-Library-Archive" (2004)
http://www.nyu.edu/tisch/preservation/program/04spring/chin-libraries.html

Week 2 (Feb 1-3) **History of Archives and Special Collections**
Read: Chap. 2 "The Nature of Archives"
Chap. 3 "Archival History and Theory"
Chap. 4 "The Uses of Archives"

Monday: Guest Speaker: Tyson Koenig, University Archives and Special Collections

Week 3 (Feb 8-10) **Archivists and Archival Institutions**
Read: Chap. 5 "Types of Archival Institution"
Society of American Archivists website: https://www2.archivists.org

Monday: explore the website for the Society of American Archivists (SAA) to become familiar with what an Archivist is/does. Pay particular attention to "The Archives Profession" tab.

Wednesday: we will be discussing the variety of archival institutions. Come to class prepared to provide 2 examples of specific archives and how you would classify them.

Remember to sign up for a Conserv-O-Gram topic by midnight on Friday, Feb. 12.

Managing an Archives/Special Collections

Week 4 (Feb 15-17) **The Archival Mission and the Mission of Your Archives**
Read: Chap. 6 "The Principles of Archival Service"

Paper #1 Due on Wednesday
In-Class Activity: What is our "Mission Statement?"

Week 5 (Feb 22-24) **Long-Range Planning and Collecting**
Read: Chap. 10 "Acquiring Archives"

In Class Activity: Creating a Collections Policy
Archival Issue Literature Review Topic due on Wednesday (Grad only)

6

Week 6 (Mar 1-3) Preservation and Protection
Read: Chap. 9 "Preserving Archives"

Preservation 101: Session 8 "Emergency Preparedness":
https://www.nedcc.org/preservation101/session-8
(be sure to read all sections of this session available on the right side of the page)

In Class Activity: Assessing likely disasters/dangers and developing a Disaster Plan

---

Week 7 (Mar 8-10) A Careful Balance: Privacy, Preservation, and Access
Read: Chap. 7 "Balancing Access and Privacy"
Chap. 12 "Making Archives Available"

Frederick J. Stielow, "Archival Security" in *Managing Archives and Archival Institutions*, James Gregory Bradshear, ed. (Chicago: Univ. of Chicago Press, 1988): 207-217 **[PDF available in Canvas]**

Wednesday: Guest Speaker: Tyson Koenig, University Archives and Special Collections: Balancing Security and Access

---

Week 8 (Mar 15-17) Digital Preservation
Read: Christopher L. Prom, ed., *Trends in Archival Practice: Digital Preservation Essentials* (Chicago, Society of American Archivists, 2016):
"Introduction" & "Module 12" (Optional reading: Module 13) **[PDF available in Canvas]**

Monday: Guest Speaker: Roxanne Dunn, University Archives and Special Collections: Working with Digital Records

Wednesday: READING DAY (no class)

Week 9 (Mar 22-24) Making the Analog Digital
Monday:
Read: NARA, "Strategy for Digitizing Archival Materials for Public Access, 2015- 2024:"
https://www.archives.gov/files/digitization/pdf/digitization-strategy-2015-2024.pdf

Missouri Historical Records Advisory Board "Digitization Guidelines for Small Historical Institutions and Repositories" (2013) **[PDF in Canvas]**

Paper #2 due on Monday

Wednesday: Group 1 only (Group 2 no class)
Meet at Special Collections & Archives for a tour and discussion of space and storage.

Read: NEDCC Preservation Leaflet 4.1: Storage Methods and Handling Practices: https://www.nedcc.org/free-resources/preservation-leaflets/4.-storage-and-handling/4.1-storage-methods-and-handling-practices

NEDCC Preservation Leaflet 4.2: Storage Furniture: A Brief Review of Current Options: https://www.nedcc.org/free-resources/preservation-leaflets/4.-storage-and-handling/4.2-storage-furniture-a-brief-review-of-current-options

Visit the website, *Preservation 101- Session 2: The Building and Environment* https://nedcc.org/preservation101/session-2 and read all sections. *Preservation 101* is a project of the Northeast Document Conservation Center, Andover, Massachusetts (NEDCC).

Processing, Storage, and Preservation

Week 10 (Mar 29-31) Identifying and Preserving Collections

Monday: Group 2 only (Group 1 no class)
Meet at Special Collections & Archives for a tour and discussion of space and storage.

Read: NEDCC Preservation Leaflet 4.1: Storage Methods and Handling Practices: https://www.nedcc.org/free-resources/preservation-leaflets/4.-storage-and-handling/4.1-storage-methods-and-handling-practices

NEDCC Preservation Leaflet 4.2: Storage Furniture: A Brief Review of Current Options: https://www.nedcc.org/free-resources/preservation-leaflets/4.-storage-and-handling/4.2-storage-furniture-a-brief-review-of-current-options

Visit the website, *Preservation 101- Session 2: The Building and Environment* https://nedcc.org/preservation101/session-2 and read all sections. *Preservation 101* is a project of the Northeast Document Conservation Center, Andover, Massachusetts (NEDCC).

Wednesday: Class time will be devoted entirely to Conserv-O-Gram Presentations

---

Week 11 (Apr 5-7) Identifying and Preserving Collections
This week's class time will be devoted entirely to Conserv-O-Gram Presentations

---

Week 12 (Apr 12-14) Processing and File Organization
Read: Chap. 11 (pp. 213-242) "Arranging and Describing Archives"

In Class Activity- Finding Aids, Arranging, and Describing

Week 13 (Apr 19-21)   Processing Collections

Read: Moravian College & Bethlehem Area Public Library Processing Manual for Archives and Special Collections: http://home.moravian.edu/public/arch/resources/manual.pdf

Society of American Archivists. *Describing Archives: A Content Standard* (2nd Edition, 2013, r. 2015). [SKIM]
E-Book: http://files.archivists.org/pubs/DACS2E-2013_v0315.pdf

Paper #3 due on Monday
Thursday's class time will be devoted entirely to processing collections for your final projects.

Week 14 (Apr 26-28)   Processing Collections
This week's class time will be devoted entirely to processing collections for your final projects.

Archival Issue Literature Review due on Monday (Grad only)

Week 15 (May 3-5) Processing Collections & Wrap Up

This week's class time will be devoted entirely to processing collections for your final projects. All students must turn in processed collection by the end of class on Thursday.

Finding Aids must be submitted in Canvas by 11:59pm on Friday, May 7.

Week 16 (May 10-12)   Finals Week

Students must collect their completed/graded archival projects between 2pm and 4pm on Monday, May 9.

Turn in Process Paper in Canvas (5-6 pages) by 11:59pm on Wednesday, May 11.

Course Syllabus
HP410/630: Introduction to Museum Studies
Southeast Missouri State University, Fall 2021
T-Th, 11-12:15, Serena 312

Eric Clements
311A Carnahan; 651–2809; eclements@semo.edu
Office Hours: 12-1 Mon. and Wed.; 1-2 Tues. and Thurs.; or by appointment

Course Objective:

This course is designed to introduce students to the elements of museum governance, operation, collections, education, and ethics through readings, discussions, and the examination of a regional museum. Students will communicate an understanding of professional issues related to governance, operation, collections management, and education at museums.

Readings:

Each student will be responsible to read required articles, as indicated on the reading list, and for weekly readings in the following required texts:

Walhimer, Museums 101, Rowman and Littlefield, 2015 (purchase).
MacLeish, The Care of Antiques and Historical Collections, AASLH Press, 1995(rental).

A Rough Course Outline:

Week One (Tuesday, 24 Aug.): Museums: definition and history.
Week Two (31 Aug.): Museum types and missions.
Week Three (7 Sep.): Museum governance and management. [Museum choice due.]
Week Four (14 Sep.): Museum operations.
Week Five (21 Sep.): Collections management theory.
Week Six (28 Sep.): Collections accession and registration methods.
Week Seven (5 Oct.): Collections preservation and interpretation. (Fall Break; no class Thurs.)
Week Eight (12 Oct.):Interpreting artifacts. [First Exam, Th., 14Oct.]
Week Nine (19 Oct.): Exhibit creation, construction, and politics.
Week Ten (26 Oct.): Education and outreach programing.
Week Eleven (2 Nov.): Development; visitor services.
Week Twelve (9 Nov.): Public relations; museum security.
Week Thirteen (16 Nov.):Museum ethics; the future; summation. [Second Exam, Th., 18 Nov.]
Week Fourteen (23 Nov.): Student presentations. (Thanksgiving; no class Thursday.)
Week Fifteen (30 Nov.): Student Presentations.
Week Sixteen(7Dec.): Student presentations.
Week Seventeen (Thursday, 16 Dec., 10am): Museum Analysis Paper due.

Assignments:

Participation:

Regular and active class participation are important elements to the success of the class and of each student. Participation will be evaluated and recorded daily, with one letter grade assigned for participation at the end of the semester.

Exams:

Two examinations will be given during the course. Each exam will cover only material considered in that section of the course. Each will be awarded one letter grade.

Museum Analysis Paper:

Each student will critique an area museum in a paper not to exceed ten pages in length. This assignment entails visiting the selected museum several times during the semester to examine its operations and gather material for analysis. The paper should consider the museum from the visitor's point of view, and also examine and evaluate the museum's governance, operations, artifacts management, exhibits, and educational programs. The paper should conclude with an assessment of the museum's present status and immediate priorities. Further ideas for this project will be provided to you in a supplemental handout.

Museum Analysis Presentation:

In our last class sessions, at a time to be assigned, each student will give a ten-to fifteen-minute presentation summarizing that student's museum visit and analysis. This presentation cannot simply be a reading of the student's analysis paper.

Grading

Each of the five assignments above will be given a letter grade. These grades will be averaged by letter to produce the final course grade.

Extra Fun for Graduate Students:

Each graduate student will complete all of the above assignments except the exams, plus write a ten to twelve-page paper on a museum topic agreed to by the instructor. This paper will receive a double-letter grade. These five grades will be averaged by letter to produce the final course grade.

Some Class Rules:

Students must operate in good faith; attempts to cut corners or manipulate the system will be frowned upon. Students guilty of academic dishonesty may fail the course. Issues will be resolved as per the appropriate university civility, accessibility or honesty policy. Questions, comments or requests regarding this course or program should be taken to your instructor. Unanswered questions or unresolved issues may be taken to Dr. Toni Alexander,chair, Department of History and Anthropology, 311E, CarnahanBuilding, or: talexander@semo.edu

Protect the Nest Classroom Guidelines:
Please refer to the Protect the Nest plan posted on the semo.edu website for current classroom guidelines and practices: semo.edu/sealerts/covid19/ protect-the-nest

HP410/630: Museum Studies Readings List

General Readings:
Walhimer, Museums 101.
MacLeish, Care of Antiques and Historical Collections.

Specific Readings (see texts, scans, and Kent Library's physical and electronic holdings):

Week One: Museums: definition and history.
Walhimer (the sections starting on pages),3, 6.

Week Two: Museum types and missions.
Walhimer, 13, 19, 33, 168, 182, 183.

Week Three: Museum governance and management.
Walhimer, 23, 39, 104, 176, 185, 199.
Harold and Susan Skramstad, "Working Model: A Mechanism for the Effective Board," *Museum News*, Jan.-Feb. 2003.

Week Four: Museum operations.
Walhimer, 79, 117.
Elizabeth Merritt, "Some Observations on Financial Stability," *Museum News*, Sep.-Oct. 2006.
Cinnamon Catlin-Legutko, "DIY Strategic Planning for Small Museums," *AASLH Technical Leaflet* no. 242, HN 63, no. 2 (Spr. 2008).
Amanda Kraus, "Extending Exhibits: Integrating the Museum Store," *Museum News*, Sep.-Oct. 2003.

Week Five: Collections management theory.
Walhimer, 49, 187.
John Simmons, "Managing Things: Crafting a Collections Policy," *Museum News*, Jan.-Feb. 2004.
James Gardner and Elizabeth Merritt, "Collections Planning: Pinning Down a Strategy," *Museum News*, Jul.-Aug. 2002.

Week Six: Collections accession and registration methods.
Walhimer, 129, 188, 205.
MacLeish, Ch. 1.Jan S. Bollard, "Computerizing Your Collections Records," *AASLH Technical Leaflet, History News* 56, no. 4 (Aut. 2001).
Kathleen Byrne, "Deaccessioning Museum Collections," *Cultural Resource Management* 23, no. 5 (2000). www.cr.nps.gov/crm

Week Seven: Collections preservation and interpretation.
Walhimer, 54.
MacLeish, Ch. 2, 3, 6, 7, 8-17.

Week Eight: Interpreting artifacts.
Walhimer, 127.Prown, "Mind in Matter: An Introduction to Material Culture Theory and Method," *Winterthur Portfolio* 17, no. 1 (Spr. 1982): 1-19.

Week Nine: Exhibit creation, construction, and politics.
Walhimer, 59, 64, 190, 209.

Alice Parman, "Exhibit Makeovers: Do-It-Yourself Exhibit Planning," *AASLH Technical Leaflet* no. 249, History News65, no. 1 (Win. 2010).
Larry Borowsky, "Telling a Story in 100 Words: Effective Label Copy," *AASLH Technical Leaflet* no. 240, History News62, no. 4 (Aut. 2007).
Susan Schreiber, "Interpreting Slavery at National Trust Sites—A Case Study in Addressing Difficult Topics," *Cultural Resource Management* 23, no. 5 (2000), www.cr.nps.gov/crm
Cary Carson, "Home from the Culture Wars, Are Museums Losing the Peace?" *History News*, 59, no. 1 (Win. 2004): 11-14.

Week Ten: Education and outreach programing.
Walhimer, 69.
Anne Grimes Rand, et al., "Families First! Rethinking Exhibits to Engage All Ages," *AASLH Technical Leaflet* no. 245, History News 64, no. 1 (Win. 2009).
Kim Fortney, "Magic Bus: Museum Programs for School Audiences: The Basics," *Museum News*, Jan.-Feb. 2011.
Mary Ellen Flannery, "It Can't Just be Fun: What Teachers Want from Field Trips," *Museum News*, Jan.-Feb.2010.
Randi Korn, "When Less Is More," Museum News, Sep.-Oct. 2010."History Bytes: Playing Games with History," *History News* 62, no. 3 (Sum. 2007): 5-6.

Week Eleven: Development; visitor services.
Walhimer, 89, 110, 122, 195, 196.
Amy Rogers Nazarov, "Tips for Fundraisers," *Museum News*, Mar.-Apr2009.
Sarah Brophy, "Is Your Site Grant Ready? How to Prepare to Attract Grants," *AASLH Technical Leaflet* no. 257. *History News* 67, no. 1 (Win. 2012).
Carol Kennis, "Save Our History: Grant Writing Tips for a Successful Application," *History News*, 61, no. 3 (Sum. 2006): 24-6. [Concerns a specific grant, but with good general principles.]
Marianne Bez and Amy Cunningham, "Membership Matters: Establishing a Vital Membership Program in Your Museum," *AASLH Technical Leaflet* no. 237, *History News* 62, no. 1 (Win. 2007).
Andrea Lewis, "Surveying Visitors, Plain and Simple," *History News* 62, no. 2 (Spr. 2007): 17-19.
Nancy Richner, et al., "All Access Pass: Making a Small Museum Disabled-Friendly," *Museum News* Jul.-Aug. 2006.
Steve Tokar, "Take a Seat," [visitor amenities] *Museum News* Sep.-Oct. 2008.

Week Twelve: Public relations; museum security.
Walhimer, 99.MacLeish, Ch. 4, 5.
Thomas Aageson, "Market Value: 5 Steps to an Effective Museum Marketing Plan," *Museum News*, July-Aug. 1999.
Stevan Layne, "An Ounce of Prevention—Worth MORE than a Pound," *AASLH Technical Leaflet* no. 253, *History News* 66, no. 1 (Win. 2011).

Week Thirteen: Museum ethics; the future; summation.
Walhimer,139, 148, 156, 164.
Elizabeth Merritt, "Running in Place, But Faster: New Changes in Museum Standards," [accreditation and accountability] *Museum News*, Mar.-Apr. 2005.
American Association of Museums, *Code of Ethics for Museums* (Washington, D.C.: AAM, 1994). [On two-hour reserve at Kent Library.]

Course Syllabus
HP420: Historic Site Administration
Southeast Missouri State University, 2012

I. Catalog Description and Credit Hours of Course:

A study of the problems, techniques and administration of historic sites. 3 credit hours.

II. Prerequisites:

Introduction to Historic Preservation or consent of instructor.

III. Purposes or Objectives of the Course:

The goal of this course is to give students academic training in the techniques of administering a historic site. During the semester, we will examine the practical issues and trends of site administration, as well as, the conceptual, philosophical, and ethical concerns. Toward this end, the course addresses: strategic planning, development, research, interpretation, economics, management, public programming, and the issues of preservation and presentation. Students will be able to think analytically about historic sites, their purposes, functions, and management.

IV. Student Learning Outcomes (Minimum of 3)

A. Students will be able to give examples of, describe, and explain the practical issues and trends of site administration, as well as, the conceptual, philosophical, and ethical concerns.
B. Students will be able to give examples of, describe, and explain the management techniques in historic site administration such as strategic planning, development, research, economics, management, public programming, and the issues of preservation and presentation.
C. Students will be able to explain and define the philosophy of interpretation of historic sites as a cultural translation from the technical language of the historian and preservationist to the ordinary language of the visitor.

V. Expectations of Students:

A. Regular class attendance
B. Participation in class discussions
C. Read all assigned materials
D. Timely completion of all reading, written, and oral assignments
E. Demonstration of mastery of course content on examinations

VI. Course Outline:

Week 1 Strategic Planning, 6 hrs.
Introduction to HP 420
Strategic Thinking: Determining Goals, Setting Objectives, and Planning

Week 2
Strategic Thinking: Determining Goals, Setting Objectives, and Planning

Week 3 Planning & Historic Site Research, 3 hrs.

Week 4 Interpretation 15 hrs.
Understanding Interpretation in Historic Sites and Organizations

Week 5 A System for Successful Interpretive Planning

Week 6
Thematic Interpretation

Week 7
Living History

Week 8
Interpretative Staff

Week 9 Heritage Education, 9 hrs.
Heritage Education

Week 10
Heritage Education

Week 11
Heritage Education

Week 12 Development, 9 hrs.
Marketing, Promotion, Fundraising, and Membership

Week 13
Marketing, Promotion, Fundraising, and Membership

Week 14
Marketing, Promotion, Fundraising, and Membership

Week 15 PastPerfect 6 hrs.
PastPerfect Museum Software Tutorial

Week 16
PastPerfect Museum Software Tutorial

Week 17 FINAL EXAM

VII. Textbooks:

William T. Alderson and Shirley Payne Low, Interpretation of Historic Sites
Hugh H. Genoways and Lynne M. Ireland, Museum Administration: An Introduction

VIII. Basis of Student Evaluation:

Heritage Education lesson plan – 20%
Evaluation Report of PastPerfect Museum Software – 20%
Mid-term exam – 30%
Final exam – 30%

There will be one mid-term and one final exam. As a team, students will also be required to develop and present a lesson plan/teaching strategy based on Heritage Education and turn in a written copy of their presentation. Individually each student will be responsible for an Evaluation Report at the conclusion of the PastPerfect Museum Software Tutorial.

Course Syllabus
HP420: Historic Site Administration
Southeast Missouri State University, Fall 2018

Dr. Joel P. Rhodes
Department of History and Anthropology
311J Carnahan
Phone: 651-2715 E-Mail: jrhodes@semo.edu
Office Hours: MWF, 11:00-12:00, Or by appointment

Class Content/Goals:

The goal of this course is to give you academic training in the techniques of administering a historic site. During the semester, we will examine the practical issues and trends of site administration, as well as, the conceptual, philosophical, and ethical concerns. Toward this end, the course addresses: strategic planning, development, research, interpretation, economics, management, public programming, and the issues of preservation and presentation. When you leave here, you should be on your way toward thinking analytically about historic sites, their purposes, functions, and management.

The curriculum of HP 420 will support three primary course goals, which are:

1. Students will be able to give examples of, describe, and explain the practical issues and trends of site administration, as well as, the conceptual, philosophical, and ethical concerns.
2. Students will be able to give examples of, describe, and explain the management techniques in historic site administration such as strategic planning, development, research, economics, management, public programming, and the issues of preservation and presentation.
3. Students will be able to explain and define the philosophy of interpretation of historic sites as a cultural translation from the technical language of the historian and preservationist to the ordinary language of the visitor.

Required Readings:

William T. Alderson and Shirley Payne Low, *Interpretation of Historic Sites*
Hugh H. Genoways and Lynne M. Ireland, *Museum Administration: An Introduction*
Additional readings and other materials will be provided periodically by the instructor.

Grading:

There will be one mid-term examination and one final exam composed entirely of short answer and essay questions. There will also be three papers based on assigned readings and AASLH Technical Leaflets. As a team, students will also be required to develop and present a lesson plan/teaching strategy based on Heritage Education and turn in a written copy of their presentation. Individually each student will be responsible for an Evaluation Report at the conclusion of the PastPerfect Museum Software Tutorial. Detailed descriptions and due dates of the paper assignments and evaluation report of PastPerfect are posted on the HP 420 website.

Class Participation – 10%
Paper #1 - 10%
Paper #2 - 10%
Paper #3 - 10%
Heritage Education lesson plan – 10%
Evaluation Report of PastPerfect Museum Software – 10%
Mid-term Exam – 20%
Final Exam – 20%

Academic Honesty:
Students are expected to familiarize themselves with the University's policy on Academic Dishonesty, including, but not limited to plagiarism, cheating, etc. Students found guilty of Academic Dishonesty may fail the assignment and/or the course and/or face penalties up to and including expulsion from the University. If you are not sure about any issue, you should ask the professor.

Tentative Class Schedule*
*This is subject to revision at the professor's discretion.

Week 1 **Strategic Planning**
August 20-24 Introduction to HP 420
Strategic Thinking: Determining Goals, Setting Objectives, and Planning
Read: McDaniel, "At Historic Houses"
Read: Genoways and Ireland, Chapters 15, 2, 3
Handouts: Kona Coffee Farm Plan
Old World Wisconsin Plan

Week 2
August 27-31 Strategic Thinking: Determining Goals, Setting Objectives, and Planning
Read: Genoways and Ireland, Chapters 4 and 13

Week 3 **Interpretive Programming**
September 3-7 Understanding Interpretation in Historic Sites and Organizations
Read: Alderson and Low, Chapters 1, 2, 3
Read: Genoways and Ireland, Chapters 12 and 14
Read: Tilden "Interpreting Our Heritage"
Handouts:
Interpretive Planning Tool Kit

Week 4
September 10-14 September 11: Glenn House Tour
Content: Planning & Historic Site Research
Read: Genoways and Ireland, Chapters 8, 9, 10
Handouts: C lermont Master Plan
Philipse Manor Hall Master Plan
Lyndhurst Furnishing Plan
Masten House Historic Structure Report
Guidelines for Preparing Historic Furnishings Reports
Landscape Maintenance Manual
Anton Sisel Furnishing Plan
Caring For Your Historic House

Paper 1 Due September 14

Week 5
September 17-21 Content: Thematic Interpretation
Read: Alderson and Low, Chapters 5 and 6

Week 6
September 24-28 — Audience: Evaluation
Read: Alderson and Low, Chapter 10
Read: Genoways and Ireland, Chapter 15

Week 7
October 1-5 — Methodology: Strategies for Successful Interpretive Planning
Read: Alderson and Low, Chapters 3 and 4
Handout: Kona Interpretive Plan

Week 8
October 8-12 — Living History
Paper 2 Due October 12

Week 9
October 15-19 — Interpretative Staff
Read: Alderson and Low, Chapters 7, 8, 9
Read: Genoways and Ireland, Chapter 7
Read: AASLH Technical Leaflet: Training Docents
Mid-term Exam

Week 10 Heritage Education
October 22-26 — Heritage Education
Read: AASLH Technical Leaflet: Historic Houses as Learning Laboratories
October 25: Bollinger Mill State Historic Site

Week 11
Oct 29-Nov 2 — Heritage Education
October 30: Lesley McDaniel, Bollinger Mill State Historic Site

Week 12
November 5-9 — Heritage Education
Read: CRM "Teaching With Historic Places"
Heritage Education Lesson Plan Due November 9

Week 13 Development
November 12-16 — Marketing, Promotion, Fundraising, and Membership
Read: Genoways and Ireland, Chapters 5, 6, 11
Read: AASLH Technical Leaflet: Nonprofit Accounting
Handouts: Financial Care and Feeding for Non-Profits
AASLH Development Bibliography
AASLH Marketing Resources
Annual Campaign Letter
Membership Sample
Membership Sample
MHC Board Development Commitment
MHC Board Responsibilities

Week 14
November 19-23 NO CLASS – THANKSGIVING, November 19-23

Week 15
November 26-30 Marketing, Promotion, Fundraising, and Membership
Read: AASLH Technical Leaflet: Fundraising
Paper 3 Due November 30

Week 16 PastPerfect
December 3-7 PastPerfect Museum Software Tutorial
Read: Alderson and Low, Chapter 10
Read: Genoways and Ireland, Chapter 15
Read: AASLH Technical Leaflet: Performance Checklist for Historical Institutions, Parts I & II

Evaluation Report of PastPerfect Due December 7

Week 17 **FINAL EXAM Due Thursday, December 13, 10am**

Course Syllabus
HP450: Problems in Historic Preservation (Museums)
Southeast Missouri State University, Spring, 2020
T-Th, 2:00–3:15, CR109

Eric Clements
311U Carnahan Building, 651-2809, eclements@semo.edu
Office Hours: 11-1, M. and W.; 12:30-1:30, T. and Th.; 4:00-5:00 Th.; or by appointment.

***Course Objectives*:**

This is an applied research course dealing with museum functions and programs. This course will be largely a lab practicum, with students working in and creating an exhibit for a local museum. Students will acquire and communicate knowledge and demonstrate skills relevant to museum work.

***Readings*:**

This course has no assigned readings, but two books have been placed on reserve at Kent Library. Both can serve as valuable references to help address some of the practical problems considered in the course. They are:

Gerald W. George and Cindy Sherrell–Leo. *Starting Right: A Basic Guide to Museum Planning* (AASLH, 2012).

Arminta Neal. *Help for the Small Museum: A Handbook of Exhibit Ideas and Methods* (Pruett Pub., 1987).

***Some Class Rules*:**

The most important rule is that students must operate in good faith. Active participation in class, and cooperation with and respect for your peers will be favorably received. Disruptive behaviors will not be permitted. Student cooperation on our group endeavors is encouraged, but students must do their own work on individual projects. Students guilty of academic dishonesty may fail the course. Issues will be resolved as per the appropriate university accessibility, civility or honesty policy. Questions, comments or requests regarding this course should be taken to the instructor. Unresolved issues involving this course may be taken to Dr. Toni Alexander, chair, Department of History and Anthropology.

***Assignments*:**

As a class, we will be offering our services to the Cape River Heritage Museum, located at the corner of Independence and Frederick. Our class will meet there unless otherwise announced. In order to further your understanding of museum practice and to assist the museum, we will complete the following assignments:

Create a Museum Exhibit: The CRHM has decided to construct an exhibit on World War I. The class as a whole will cooperate to research, design, assemble, and construct this exhibit to fit the space designated. As a member of one of the exhibit creation teams, you will be evaluated individually on the quality and quantity of your participation.

Museum Exhibit Grade: Once assembled, the exhibit itself will be graded. It will be evaluated on the quality of its construction, content, and visual appeal. This is to say that you will be judged as a group on your results.

Give an Exhibit Proposal Presentation: During the last two weeks of the course you will give an individual exhibit design presentation, not to exceed fifteen minutes. Your exhibit idea must fit the CRHM's mission statement and concern some aspect of Cape Girardeau's history. The total cost of your proposed exhibit cannot exceed five hundred dollars. (See Neal for ideas on creating and presenting exhibit proposals.)

Create a Written Exhibit Proposal: You will write an individual exhibit proposal, not to exceed ten pages, detailing the ideas offered in your exhibit presentation. Your proposal will discuss your exhibit's theme and main ideas, describe its arrangement and construction, and contain a budget.

Participate in Museum Activities: You will receive marks for participation in the course through your general work in operating the museum and in completing other museum tasks as assigned. Your participation will be evaluated at the end of the semester to produce your participation grade.

These assignments will produce a total of five letter grades, which will be averaged by letter to produce your final course grade.

***Goals*:**

Goal 1: Create and construct an exhibit for the museum.
Goal 2: Create an individual museum exhibit proposal for the museum.
Goal 3: Other duties as determined.
Goal 4: Operate the museum as docent staff.

***Deadlines*:**

Museum Exhibit due: Tues., 6 March.
Individual Exhibit Presentations: Weeks of 28 Apr. and 5 May.
Individual Written Exhibit Proposal: Tues., 12 May.
Museum Exhibit Grade awarded: Fri., 13 March.
Museum activities participation grade awarded: Fri., 8 May.

Course Schedule:

Week One (Tuesday, 21 Jan.): Course introduction; initial museum visit and planning.
Week Two (28 Jan.): Goals 1 and 3; CRHM and archives, sections.
Week Three (4 Feb.): Goals 1 and 3; CRHM and archives, sections.
Week Four (11 Feb.): Goals 1 and 3; CRHM and archives, sections.
Week Five (18 Feb.): Goals 1 and 3; CRHM and archives, sections.
Week Six (25 Feb.): Goals 1 and 3; CRHM, sections.
Week Seven (3 Mar.): Goals 1 and 3; CRHM, sections.
Week Eight (10 Mar.): Goals 1 and 3: CRHM, sections.
Week Nine (17 Mar.): Spring Break.
Week Ten (24 Mar.): Goals 2, 3 and 4; no scheduled meetings.
Week Eleven (31 Mar.): Goals 2, 3 and 4; CRHM, sections.
Week Twelve (7 Apr.): Goals 2, 3 and 4; CRHM, sections.
Week Thirteen (14 Apr.): Goals 2, 3 and 4; CRHM, sections.
Week Fourteen (21 Apr.): Goals 2, 3 and 4; CRHM, sections.
Week Fifteen (28 Apr.): Goal 2; Individual exhibit presentations; CR109, all.
Week Sixteen (5 May): Goal 2; Individual exhibit presentations; CR109, all.
Week Seventeen (Tues., 12 May, 2pm): Written Exhibit Proposal due.

Course Syllabus
HP450/650: Problems in Historic Site Administration
Southeast Missouri State University, Spring 2021

Dr. Joel P. Rhodes, Department of History and Anthropology, 311J Carnahan
Phone: 651-2715 E-Mail: jrhodes@semo.edu
Office Hours: MWF, 11:00-12:00, or by appointment

Class Content/Goals:

The goal of this course is to give you the opportunity to apply academic training in the administration of a historic site to practical situations outside the traditional classroom setting. This semester the course explores performed interpretation with an emphasis on the experiential application of living history. In collaboration with Old Town Cape, students will develop a Talking Tombstone Tour at Old Lorimier Cemetery.

To do this we will examine living history as a philosophy, movement, technique, and educational interpretative tool, and will develop an understanding of how living history can create understanding and meaning and can convey an historic site's significance to a contemporary audience. The assigned readings, lectures, and discussions will facilitate student's active participation in the planning, development, and implementation of living history programming at local historic sites and events.

The curriculum of HP 450/650 will support three primary course goals, which are:

1. Students will be able to apply academic training in the experiential administration of historic sites, interpretation of material culture, and the preservation of historical documents, objects, buildings, and landscapes using practical situations outside the traditional classroom setting.
2. Students will be able to experientially carry out historic preservation project planning and management.
3. Students will be able to communicate, in written and oral forms, using the technical language of the historic preservation professional.

Readings:

Joyce M. Thierer, Telling History: A Manual for Performers and Presenters of First-Person Narrative. Nashville, TV: Alta Mira Press, 2010.
Additional readings, materials, and bibliographies will be provided periodically by the professor.

Communication:

The best way to contact me during the semester is via e-mail. I will contact the class through Announcements in Canvas so please check those regularly. Even if you turn off Announcements, you are still responsible for material in those emails.

Questions, comments or requests regarding this course or program should be taken to your professor. Unanswered questions or unresolved issues involving this class may be taken to Dr. Toni Alexander, Chair of the Department of History and Anthropology.

Academic Honesty:

Students are expected to familiarize themselves with the University's policy on Academic Dishonesty, including, but not limited to plagiarism, cheating, etc. Students found guilty of Academic

Dishonesty may fail the assignment and/or the course and/or face penalties up to and including expulsion from the University. If you are not sure about any issue, you should ask the professor.
Basis of Student Evaluation:

Class Participation (10%)

Active and engaged participation in class discussions, presentations, exercises, and rehearsals.

Character Worldview Research Paper (30%) - Due April 23

Working individually, students will research and write a 15-20-page paper addressing the "Five Spheres of Knowledge" which provide the framework for developing "character worldview."

Living History (60%)

Working as a team, students will provide one (1) copy of a Talking Tombstone Program (Due April 30) to include, but not limited to:

1. Identification of any interpretive goals and themes
2. Historical background and context of setting
3. Interpretive storyline
4. Tour Format
5. Tour Route Map for visitors
6. Individual First-Person Narrative Scripts

Graduate Students will additionally:

1. Facilitate the compilation of the Talking Tombstone Program
2. Develop the Tour Route Map for visitors - Due April 30
3. Participate as liaisons with Old Town Cape for event promotion

Working individually, students will contribute one (1) copy of a First-Person Narrative Script to the Talking Tombstone Program to include, but not limited to:

1. Identification of the historic figure/character
2. Narrative Script
3. Outline of appropriate costuming

Working as a team, students will participate in an approximately two-hour Talking Tombstone Tour at Old Lorimier Cemetery on Friday, May 7, 6:00-8:00pm

Working as a team, students will participate in an approximately two-hour virtual Talking Tombstone Tour at Old Lorimier Cemetery

Working individually, students will contribute a 5-10-minute costumed monologue and Q&A in character and as a scholar in the live and virtual Talking Tombstone Tours.

Class Schedule*

*This is subject to revision at the professor's discretion.

Week 1 January 25-29 Philosophy of Historic Interpretation

Introduction and Historic Interpretation

Week 2 February 1-5 Living History and Historical Performance

Performed Interpretation: Historical Performance and Living History

Read: 1. Thierer, Chapter 1
2. Jay Anderson, "Living History," in Jay Anderson, A Living History Reader (1991).

Week 3 February 8-12 Planning for Performed Interpretation
Determine Interpretive Themes
Character Development
Read: 1. Thierer, Chapter 2
2. John Fortier, "Louisbourg: Managing a Moment in Time," in Jay Anderson, A Living History Reader (1991).

Week 4 February 15-19
Read: 1. Thierer, Chapter 3
2. James Deetz, "The Reality of the Pilgrim Fathers," in Jay Anderson, A Living History Reader (1991).

Week 5 February 22-26 General Training in Strategies and Techniques for Interaction
No Class – Reading Day, Tuesday, February 23
Acting Workshop
Read: 1. Thierer, Chapter 4

Week 6 March 1-5 General Training in Strategies and Techniques for Interaction
Read: 1. Andrew H. Baker and Warren Leon, "Old Sturbridge Village Introduces Social Conflict into Its Interpretation," in Jay Anderson, A Living History Reader (1991).

Week 7 March 8-12 General Training in Strategies and Techniques for Interaction
Read: 1. Thierer, Chapter 5
2. Tracey Linton Craig, "Retreat Into History," in Jay Anderson, A Living History Reader (1991).

Week 8 March 15-19 General Training in Strategies and Techniques for Interaction
Read: 1. Thierer, Chapters 6, 7, 8

Week 9 March 22-26
Focused Training Sessions

Week 10 March 29-April 2
Focused Training Sessions

Week 11 April 5-9
Focused Training Sessions

Week 12 April 12-16
Focused Training Sessions

Week 13 April 19-23
Rehearsals at Old Lorimier Cemetery
Character Worldview Research Paper Due April 23

Week 14 April 26-30
Rehearsals at Old Lorimier Cemetery
Talking Tombstone Program Due April 30

Week 15 May 3-7
Record Virtual Talking Tombstone Tour
Showtime, Friday, May 7, 6:00-8:00pm

[CLASSROOM GUIDELINES FOR MINIMIZING THE RISK OF COVID-19]

• Practice social distancing (six feet of spacing), wear facial coverings, and follow proper prevention hygiene, such as washing your hands frequently and using alcohol-based (at least 60% alcohol) hand sanitizer when soap and water are not available.

• The wearing of a face covering is a University safety requirement under the current conditions as well as a Cape Girardeau County Emergency Order pursuant to RSMo 192.300. We hope that through modeling and a positive and encouraging environment, all will do their part. If a student refuses to wear a face covering, it would be handled in the same manner as if they refused to comply with a course requirement or adhere to the Code of Student Conduct. An instructor may ask a student to leave a class and an employee may ask a person to leave a meeting or event if they do not wish to comply with the Face Coverings and Social Distance Guidelines.

• In the event that a student does not comply with this guidance and does not leave the space, faculty/staff should send the student's name and email address or Southeast ID to deanofstudents@semo.edu. If a student refuses to leave a classroom or other University facility upon request, and/or becomes disruptive, the Department of Public Safety can be contacted for assistance.

• If you are not able to wear face coverings due to health or ADA concerns, contact the office of Disability Services at ds@semo.edu or 573-651-5927 for information on how to obtain an exemption. Affected individuals should also consult with their physicians before wearing a face covering.

• Avoid sharing electronic devices, books, pens, and other learning aids unless thoroughly disinfected between users.

• Students are encouraged to clean and disinfect before and after class, any contact surfaces such as chairs, desks, and equipment utilized for the class.

• Appropriate signage will be displayed in the classroom or adjacent areas on protocols to follow that protect our community members during classes. Please become familiar with this information and follow it in our shared community.

• The seating in your classroom may be positioned to maintain social distancing or signs may be posted with seating direction. Do not move seating or signs unless instructed to do so by your instructor.

• Social Distancing minimizes risk. When in buildings, hallways and entering or exiting buildings or classrooms, maintain at least six feet of social distancing space and avoid gathering in groups.

• To minimize risk and to follow the face covering guidelines, there should be no eating in the classrooms or indoor instructional spaces. Drinks are permissible provided face coverings are removed only while drinking and are properly worn when not drinking. If other guidelines regarding no food or drink are posted or conveyed, those guidelines must be followed. Medical necessities for food should be communicated to the instructor and are permissible. Students should minimize the time without a face covering as much as possible.

Course Syllabus
HP450/650: Problems in Historic Preservation: Archives & Special Collections
Southeast Missouri State University, Fall 2021
T/Th 9:30-10:45, Carnahan 109

Instructor: Dr. Lily Santoro
Office: Carnahan 311C; Email:lsantoro@semo.edu
Hours:Tues.2-4:30& Wed.12-3:30
Phone:573.651.2831 (during office hours only)

Course Description:
Students will have the opportunity to apply academic knowledge to practical situations by carrying out group projects outside the traditional classroom setting. The specific focus of this course will be archives management.

Course Learning Outcomes:
1. Students will be able to demonstrate knowledge of relevant issues in the field of archives, museums, historic sites or built environment preservation.
2. Students will be able to communicate, in written and oral forms, using the technical language of the historic preservation professional.
3. Students will be able to demonstrate knowledge and skills relevant to the preservation, interpretation, presentation, management or administration of historical documents, objects, buildings, landscapes, or historic sites through successful completion of a project.

Course Requirements:

PARTICIPATION & IN CLASS ACTIVITIES:
Students are expected to attend class and participate in class discussion and in-class activities. Everyone is responsible for all of the assigned readings. All readings should be completed before class on the day or week assigned. There will be several in-class activities throughout the semester that count towards this portion of your grade. Not attending class will result in missed participation opportunities and will hurt your grade. Participation grades are dependent upon active and appropriate participation in discussion and careful completion of in-class activities (silent, uncooperative, or destructive attendance does not earn credit). Participation and in-class activities 15% for both HP450 and HP650.

INDIVIDUAL ASSIGNMENTS:PAPERS
Students will write 3 papers (3 pages each).
- -Paper 1: Environment Survey/Assessment
- -Paper 2: Rare Book Care Proposal
- -Paper 3: Arrangement & Description Proposal

Instructions for completing each paper will be posted in Canvas. Follow the instructions for each paper closely. Late papers will not be accepted. Papers that violate the University's academic honesty policy will earn a zero. Each paper is worth 15% for the course grade in HP450. It is worth 10% for grad students enrolled in HP650.

CLASS PROJECT:MAAPS DIGITIZATION PROJECT

This semester, we will be completing a pilot digitization project for the Malden Army Airfield Preservation Steering Committee (MAAPS). This project will involve planning, digitizing, and managing digital (and physical) records for the organization. Grading will be based upon the following components:-Project Planning-Digitization-Metadata/Record Management in Preservica-Digitization Workflow Manual Students will work as a class for the planning phase of this project and work in teams of 3-4to complete the rest of the digitization project. As a class, students will compile the information from all steps of the project to create a coherent digital collection in Preservica and digital workflow manual to be bound and presented to MAAPS at the end of the semester. Each component of the project will be worth 10% of the course grade for both HP450 and HP650.

GRADUATE PROJECT:

Students enrolled in HP650 (for graduate credit) will work in teams of four to create an online exhibit, highlighting an archival collection. Teams and topics will be assigned in Week 4. Each group will meet with Dr. Santoro at least three times throughout the semester. The project is worth 15% of the course grade in HP650.

How Grades are Determined:

Students will be evaluated on active participation in all class activities, written reports, and performance on project work.

Assignment	HP450	HP650
Participation & Activities	15%	15%
Paper 1	15%	10%
Paper 2	15%	10%
Paper 3	15%	10%
Digital Project Planning	10%	10%
Digitization	10%	10%
Metadata/Record Mgmt	10%	10%
Digitization Manual	10%	10%
Graduate Project	N/A	15%
Total	100%	100%

HP450 COURSE GRADE SCALE:

A : 90-100
B: 80-89.99
C: 70-79.99
D: 60-69.99
F: 0-59.99

HP650 COURSE GRADE SCALE:

A : 90-100
B: 80-89.99
C: 70-79.99
F: 0-69

Required Readings:

Students are expected to keep up with all assigned readings listed in the syllabus and complete any additional reading assignments given in class or announced in Canvas. Readings are listed for each week in the Course Schedule. Students are expected to have read the assignment for the week before attending class.

TEXTBOOK:
The required textbook for this course is:
Gregory S. Hunter, Developing and Maintaining Practical Archives, 3rdEd. (2020).

The required book for this course is available through Textbook Rental.

ADDITIONAL READINGS:
There will also be additional readings listed in the Course Schedule and made available on Canvas: https://semo.instructure.com

Learning Environment Expectations:

CLASSROOM DISCUSSION & CIVILITY:
All students are expected to be active participants in class, and join in polite discussion (no name-calling, slurs, or talking over each other). All students must exhibit respect for their classmates (and their opinions, questions, or answers) at all times. Students should be familiar with University Civility policies available here: http://www.semo.edu/stuconduct/general-info.html

FOOD AND BEVERAGES:
Because much of our class-time will be devoted to a hands-on archival projects and activities with real historical documents/records, it is imperative that students adhere to best practices in the classroom. For this reason, food and beverages will not be allowed in the classroom. This is considered best practice and will be strictly enforced in this class. This is for the protection of the documents. No exceptions.

PROTECT THE NEST CLASSROOM GUIDELINES:
Please refer to the Protect the Nest plan posted on the semo.edu website for current classroom guidelines and practices. The direct link is: Protect the Nest -Southeast Missouri State University (semo.edu) or https://semo.edu/sealerts/covid19/protect-the-nest

ELECTRONIC DEVICES:
While accidents happen (cell phones ring, etc.),text messaging, video games, facebook, "google-ing," etc. are NOT allowed during class. Unless specifically instructed to do so, students may not use their phones, ipods, apple watch, or other "wearables" during class. If you are using these devices during class, you will be counted as absent for the day (negating any participation points earned that day).Use of laptops, tablets or other electronics must be strictly limited classroom appropriate work for this course during class time. Students who violate this policy will not be allowed to use these devices in class.

ACADEMIC HONESTY & PLAGIARISM:
Southeast Missouri State University expects all students, faculty and staff to operate in an honest and ethical manner. Academic dishonesty is a very serious offense because it undermines the value of your education and the education of others. Students who engage in academic dishonesty face significant penalties. Forms of academic dishonesty include, but are not limited to, plagiarism, cheating, contract cheating, misrepresentation, and other actions you take. Some of these are defined below:

•Plagiarism means passing off someone else's work as your own, whether it is intentional or unintentional.
•Cheating includes copying from another person or source of information to meet the requirements of a task.
•Contract cheating is paying someone else or a company to do your work.
•Misrepresentation means you are posing as someone else or someone else is posing as you to complete a task.
•Collusion means working with one or more people to cheat. If you help someone cheat or plagiarize you will face the same penalties.

For more information, visit the Responsible Redhawks Guide Student Code of Conductor the Faculty Handbook(See Section (D) on Academic Honesty).

All students who violate the University's academic honesty policies in this class will receive a zero for the assignment (without the opportunity to make it up) and may be reported to Student Conduct. Make sure the language and analysis you use in your written assignments is your own. If you are using someone else's ideas or insights, provide accurate citations and quotation marks. If you are unsure, consult with the professor before turning in the assignment. It is always best to err on the side of caution when dealing with questions of academic honesty. Plagiarism and academic dishonesty will not be tolerated in this course. No exceptions will be granted.

TECHNOLOGY:
All written assignments must be typed, using a word-processing program, such as MS Word. Students will have the opportunity, but will not be required, to submit their papers electronically in Moodle. Students will be required to access some of the reading material for the course on the internet and must have access to software that will read PDFs and graphic-heavy websites. Students will also be required to work with digital technology(including scanning, video, and data management software) for multiple projects and activities in and out of the classroom. Students who anticipate difficulty in mastering any of the technological components of the course are responsible for seeking assistance in a timely manner. The primary mode of extra-classroom communication for this course will be through University-provided email. All students are expected to check their @semo.edu email accounts for updates and messages regularly (typically once a day).

LATE WORK/MISSED ACTIVITIES:
In the majority of cases, presentations, hands-on activities, and other in-class exercises cannot be made up and late papers are not generally accepted. In extremely unusual circumstances (excused absences with written verification from the Dean of Students), a student may need to make up missed work. Should this happen, the student is responsible for contacting the professor and making arrangements for a make-up or alternate assignment as soon as possible. The professor will not make arrangements with students who have not made contact within three days of the due date.

ACCESSIBILITY:
Southeast Missouri State University and the Office of Accessibility Services are committed to making every reasonable educational accommodation for students who identify as people with disabilities. Many services and accommodations which aid a student's educational experience are available for students with various disabilities. Students are responsible for contacting The Office of Accessibility Services to register and access accommodations. Accommodations are implemented on a case-by-case basis. For more information, visit the Accessibility Services page

or contact Accessibility Services by phone at 573-651-5927.Communication Expectations: Questions, comments or requests regarding this course or program should be taken to your instructor, Dr. Lily Santoro. Unanswered questions or unresolved issues involving this class may be taken to Dr. Toni Alexander, Chair, Department of History & Anthropology, who can be contacted at 573-651-2478.

OFFICE HOURS: Drop-in office hours will be held in Carnahan 311C on Tuesday 2-4:30 and Wednesdays 12-3:30. If you prefer to talk by phone, you may also call during office hours: (573) 651-2831.Students who prefer a video chat may meet with the instructor on Zoom by appointment. Appointments can be arranged by sending an email to lsantoro@semo.edu

DIRECT COMMUNICATION WITH INSTRUCTOR: Dr. Santoro's primary mode of personal communication is email (lsantoro@semo.edu). Typically, she will respond to emails within 24 hours, Mon-Fri, and within 48 hours on weekends and during University holidays. Because class related communication with students is only authorized via official SEMO email addresses, students must use their @semo.edu email to contact Dr. Santoro and should check that email box daily.

CLASS ANNOUNCEMENTS: Class-wide announcements will be posted in the Announcements Forum on our Canvas page. Most students receive these announcements in their @semo.edu email, but you should check that forum regularly to be sure you are caught up.

CONFIDENTIALITY AND MANDATORY REPORTING: I will keep information you share with me confidential to the best of my ability, but as a professor I am legally required to share information about sexual misconduct and crimes I learn about to make our campus and community safe for everyone.

Course Schedule:
Note: Course schedule may be subject to change. Any changes will be announced in class, Moodle, and email.

Week 1 (Aug. 24 -26) THE ARCHIVAL PROFESSION

Read: Developing and Maintaining Practical Archives, Chapters 1 & 12

Week 2 (Aug. 31-Sept. 2) ARCHIVAL COLLECTIONS & INSTITUTIONS

Read: Developing and Maintaining Practical Archives,Chapters2 & 3

Week 3 (Sept. 7-9) APPRAISING & COLLECTING

Read: Developing and Maintaining Practical Archives,Chapters4 & 5

Week 4 (Sept. 14-16) ARRANGEMENT & DESCRIPTION

Read: Developing and Maintaining Practical Archives, Chapters 6& 7

Week 5 (Sept. 21-23)THE ACT OF PROCESSING A COLLECTION

Read: Manuscript and Archival Records Processing Manual(Arizona State University Archives and Special Collections) https://www2.archivists.org/sites/all/files/asu_processing_manual.pdfProcessing Manual for Archival and Manuscript Collections(University of Massachusetts at Boston) https://scholarworks.umb.edu/cgi/viewcontent.cgi?article=1028&context=hlpubs

Week 6 (Sept. 28-30) DIGITIZATION PROJECT INTRO

Read: Moving Theory into Practice: Digital Imaging Tutorial (Cornell University Library) http://preservationtutorial.library.cornell.edu/contents.html
Review Website: Santa Barbara Historical Museum Digital Collections (Preservica showcase) https://t08a9146c20e4db40.starter1ua.preservica.com
We will meet at Special Collections & Archives on Thursday of this week.

Week 7 (Oct. 5-7) ACCESS AND REFERENCE
Read: Developing and Maintaining Practical Archives, Chapter 9
Wendy Duff and Allyson Fox, "You're a guide rather than an expert': Archival reference from an archivist's point of view," Journal of the Society of Archivists(Oct 2006), Vol. 27 Issue 2, 129-153. Available at: https://library.semo.edu:2443/login?url=http://search.ebscohost.com/login.aspx?direct=true&db=a9h&AN=23462859&site=ehost-live
Note: Class meets Monday only this week .The university is closed for Fall Break Oct. 7-8

Week 8 (Oct. 12-14) THE PRESERVATION OF ARCHIVAL RECORDS
Read: Developing and Maintaining Practical Archives, Chapter 8
Visit the website, Preservation 101: Session 2 The Building and Environment https://www.nedcc.org/preservation101/session-2 and read all sections listed under Building and Environment on the right side of the page.
*Preservation 101 is a project of the Northeast Document Conservation Center, Andover, Massachusetts (NEDCC).
Grad Project Plan due Thursday, Oct. 14

Week 9 (Oct. 19-21) CLEANING, & STABILIZING ANALOG COLLECTIONS
Read: Visit the website, Preservation 101: Session 4 Caring for Paper Collectionshttps://www.nedcc.org/preservation101/session-4and read all sections under Caring for Paper Collections on the right side of the page.*Preservation 101 is a project of the Northeast Document Conservation Center, Andover, Massachusetts (NEDCC).NEDCC Preservation Leaflet: 7.2 Surface Cleaning of Paper https://www.nedcc.org/assets/media/documents/Preservation Leaflets/7_4_enclosures_2018.pdf
Paper #1 (Environmental Survey) due Thursday, Oct. 21

Week 10 (Oct. 26-28) CARE & STORAGE FOR BOOKS & BOUND MATERIALS
Read: NEDCC Preservation Leaflets:

4.1 Storage Methods and Handling Practices http://www.nedcc.org/free-resources/preservation-leaflets/4.-storage-and-handling/4.1-storage-methods-and-handling-practices
4.2 Storage Furniture: A Brief Review of Current Options http://www.nedcc.org/free-resources/preservation-leaflets/4.-storage-and-handling/4.2-storage-furniture-a-brief-review-of-current-options
4.3 Cleaning Books and Shelveshttp://www.nedcc.org/free-resources/preservation-leaflets/4.-storage-and-handling/4.3-cleaning-books-and-shelves

Week 11 (Nov. 2-4) RARE BOOK & BOUND MATERIALS CARE
Read: NEDCC Preservation Leaflets:

4.7 The Book Shoe: Description and Useshttp://www.nedcc.org/free-resources/preservation-leaflets/4.-storage-and-handling/4.7-the-book-shoe-description-and-uses
4.9 Storage Solutions for Oversized Paper Artifacts http://www.nedcc.org/free-resources/preservation-leaflets/4.-storage-and-handling/4.9-storage-solutions-for-oversized-paper-artifacts
7.4Custom Protective Enclosures
https://www.nedcc.org/assets/media/documents/Preservation Leaflets/7_4_enclosures_2018.pdf

Paper #2 (Rare Book Care Proposal) due Thursday, Nov. 4

Week 12 (Nov. 9-11) ARRANGEMENT AND DESCRIPTION REVISITED

Read: Mark A. Greene and Dennis Meissner, “More Product, Less Process: Revamping Traditional Archival Processing,” The American Archivist68 (Fall/Winter 2005): 208-265. Available at https://americanarchivist.org/doi/pdf/10.17723/aarc.68.2.c741823776k65863

We will meet at Special Collections & Archives on Tuesday of this week.

Week 13 (Nov. 16-18) WORKING ON PROJECTS

Class does not meet in person. Dr. Santoro is available in her office for questions/consultation during our usual class time and office hours.

Paper #3(Arrangement & Description Proposal) due Tuesday, Nov. 16

Week 14 (Nov. 23 -25) CHECKING IN / THANKSGIVING BREAK

Dr. Santoro will meet with each group to discuss progress on projects. Please schedule a meeting time with her in Canvas.

Note: Meetings will be on Monday& Tuesday only. The university is closed Nov. 24-26

Week 15 (Nov. 30-Dec. 2) WORKING ON PROJECTS

Class does not meet in person. Dr. Santoro is available in her office for questions/consultation during our usual class time and office hours. Grad Project due Thursday, Dec. 2

Week 16 (Dec. 7-9) WORKING ON PROJECTS

Class does not meet in person. Dr. Santoro is available in her office for questions/consultation during our usual class time and office hours.

Finals Week(Dec. 16):

Class Project and peer evaluations are due by 5pm on Dec. 16 .

Course Syllabus
HP552: Historic Preservation Field School
Southeast Missouri State University, Summer 2021

Instructors:
Dr. Steven J. Hoffman
Department of History
311T Carnahan Hall
email: shoffman@semo.edu

Office Telephone: (573) 651-2808
Home Telephone: (573) 335-2820
Cell Phone: (573) 225-8393

Ms. Donna Rausch, Site Administrator
(and famous Southeast HP graduate)
Mr. Jim Baker
Felix Vallé House State Historic Site
P.O. Box 89
Ste. Genevieve, MO

Office Telephone: (573) 883-7102
felix.valle.state.historic.site@dnr.mo.gov

Course Description:

A workshop for graduate, undergraduate and continuing education students involving intensive field experience in documenting, interpreting and preserving the built environment and the historic landscape.

Students will receive hands-on field experience in documenting, preserving and interpreting the built environment and the historic landscape, including work with standing structures and historic sites. Authentic 18th and early 19th century buildings in the Felix Valle State Historic Site will serve as learning laboratories, providing hands-on investigation of Ste. Genevieve's rare vertical log structures. The goal of the Field School is to acquaint students with the broad array of skills and techniques affecting historic preservation issues and historic site management. Students will learn to gather and record architectural data, utilizing library and archival collections in researching history as tools for understanding historic structures. (3 credit hours)

Objectives of the Course:

1. To provide field experience in the documentation and preservation of the built environment.
2. To familiarize students with various methods of gathering data and understanding past cultures, including material culture and historic archaeology.
3. To deepen students' understanding of the historic landscape.
4. To strengthen students' skills in field observation and historical research.
5. To develop a basic understanding of the operation of historic sites.
6. To broaden student understanding of past human behavior from a multi-disciplinary and multi-cultural perspective.

Student Learning Objectives:

1. Students will be able to demonstrate knowledge and skills relevant to the documentation of historic buildings, landscapes, or sites.
2. Students will be able to demonstrate knowledge and skills relevant to the preservation of historic buildings, landscapes, or sites.
3. Students will be able to demonstrate knowledge and skills relevant to the interpretation of historic buildings, landscapes, or sites.

Expectations of students:
Students are expected to: attend all sessions, read assigned materials, participate in all field trips and field studies, conduct themselves as professionals, participate fully in group projects, and complete any written assignments. In addition, graduate students will have additional readings and be responsible for leading portions of discussion.

Proposed Schedule:
Meet daily, 9:00 a.m. – 5:00 p.m.(Monday-Saturday, May 17-22, Monday-Friday, May 24-28 at the Felix Valle State Historic Site, 198 Merchant St., in the Education room/Stone Gallery behind the Shaw House, Ste. Genevieve, Missouri.

Monday, May 17
Introduction and Orientation
Planning Process for Historic Sites
Presentation of Interpretive Plan
Tour: Felix Vallé House State Historic Site

Tuesday, May 18
Assignment for Tuesday: Read Interpretative Plan; Read Invitation to Vernacular Architecture, Introduction and Chapter 1 (pp. xiii-18)
Discuss Felix Valle interpretative plan
Tour:
11-12 Welcome Center and JB Vallé garden and house
12-1 Lunch
1-2 Interp Program at the Amoureux House.
2-3 Carrie Miller, Denver Service Center, discussing Foundation Documents and her career
3-4 Ste Genevieve NHP panel discussion on careers/Q&A

Wednesday, May 19
Assignment for Wednesday: Read Invitation to Vernacular Architecture, Chapter 2 (pp. 19-43)
Document Transcription
Sanborn Map exercise
Lunch
Bolduc House (1/1:30)
Guibourd-Vallé House
Time permitting: Ste. Genevieve Memorial Cemetery

Thursday, May 20
Assignment for Thursday: Read Invitation to Vernacular Architecture, Chapter 3 (pp. 45-62)
Discussion of Readings
Photography Workshop Part 1: 10:30-12:30
Lunch
Part 2: field work: location: Bob Brown House, 498 Merchant Street): 1:30-2:45
3:00 Amoureux HSR presentation
Work on Projects (time permitting)

Friday, May 21
Assignment for Wednesday: Read Invitation to Vernacular Architecture, Chapter 4 (pp. 65-81)
Archeology at Pierre Dorlac House (Richard Young, famous Southeast HP graduate)

Saturday, May 22
Measured Drawings (Richard Young, famous Southeast HP graduate)
Location: (Location: Bob Brown House, 498 Merchant Street)

Monday, May 24
Discussion of Invitation to Vernacular Architecture
Ste. Genevieve Catholic Church
Ste. Genevieve Museum
Lalumundier House
Guibourd-Vallé House

Tuesday, May 25
Heritage Education project: Paper Marbling
Lunch: Graduate Student Additional Readings discussion
Norman Truss races. Time: TBA
Field Assignment: Find 2-3 variations of a mortise and tenon joint at the Guibourd-Vallé house and sketch them. Time: TBA
Tour: Bequette Ribault (4 pm)
Work on Projects (time permitting)

Wednesday, May 26
Window Repair Workshop
Kern House (11 am?)
Bousillage workshop

Thursday, May 27
Timber framing workshop (Jesse Francis, St. Louis County Parks Department and famous Southeast HP graduate)
Work on Projects
"What the Tour Guides Won't Tell You" walking tour

Friday, May 28
Individuals present transcribed documents
Groups Present Interpretive Plans (Felix Valle Wash House Interpretative Plan and Dr. Shaw Office Interpretative Plan)
BBQ lunch – Lion's Club Park, Market Street

Textbooks
Thomas Carter, Elizabeth Collins Cromley, *Invitation to Vernacular Architecture: A Guide to the Study of Ordinary Buildings and Landscapes* (University of Tennessee Press, October 24, 2005)
Various additional readings will be assigned to enhance student understanding of observations made in the field and to relate to the particular projects being pursued.

Basis for Student Evaluation
Grades for undergraduates will be based on performance in field exercises (50%), written assignments (30%), and class participation (20%).

Graduate students will be evaluated on field exercises (50%), written assignments (20%), and class participation (30%), including leading portions of discussion of additional readings.

Course Syllabus
HP580: History of American Building Materials and Techniques
Southeast Missouri State University, Spring 2021
T/Th 11:00-12:15, CR109

Dr. Steven J. Hoffman
Department of History
311 T Carnahan Hall
Office: (573) 651-2808
Cell: (573) 225-8393

Office Hours: T/Th 1:00-2:30
and by appointment
email: shoffman@semo.edu

www.facebook.com/hoffman.steven

Teaching Assistant:
Madelyn Brown
email: mpbrown2s@semo.edu
Cell: (636) 696-3722

Note: office hours will be conducted remotely. Please call or email to set up a remote session via Zoom. You should expect to receive a response to emails within 24-48 hours.

Course Description:
A survey of traditional construction materials and methods used in buildings and structures in the United States from the late sixteenth century to the recent past. Topics include: history of various construction materials and techniques, performance and behavior of materials, common building pathologies and remediation techniques, building fabric, and ornamental and decorative finishes. (3 hours)

Course Objectives:
A. To introduce students to the theory and practice of conservation, restoration, and rehabilitation of historic structures, specifically as they relate to the appropriate treatment of historic building materials.
B. To familiarize students with the manufacture, use, deterioration, repair, and appropriate replacement of traditional building materials employed in historic building construction.
C. To provide students with the opportunity to apply material learned in class to the forensic analysis of older buildings, structures and sites.

Student Learning Outcomes:
A. Students will be able to identify a variety of historic building materials.
B. Students will be able to identify deterioration in at least one architectural feature, diagnose the probable cause (or causes) of the deterioration and offer recommended steps for remediation.
C. Students will be able to assess the architectural and material condition of a historic building using established procedures and criteria in the historic preservation field.

COVID-19 Updates:
As part of our COVID-19 accommodations we will be dividing the class into two cohorts: We will have a Tuesday cohort (Group A) and a Thursday cohort (Group B). The two cohorts will alternate class and remote (synchronous) attendance. In addition, there will be several field trips over the course of the semester where the entire class can meet together safely since we will be outside, masked, observing social distancing protocols, and are a fairly small group. These trips will be identified well in advance to facilitate scheduling.

Remote participation: Students will be expected to participate remotely using Zoom during regularly scheduled class times. Students should, to the extent feasible, keep their camera on during class to help foster a sense of class community. Please let the professor know in advance if there is a reason for not having the camera on during any particular class session. Class sessions will be recorded and posted in the Office365 Teams/Stream folder for the class.

Classroom Guideline for Minimizing the Risk of COVID-19 **[see above, page XX]**

Expectation of Students:
Students will be expected to attend all classes—including field trips, do all required readings, demonstrate an acceptable degree of understanding on written examinations and/or papers, and participate in individual, group and class projects.

Textbook(s) and/or Other Required Materials or Equipment:

Textbook Rental:

Young, Robert A. Historic Preservation Technology: A Primer. Hoboken, NJ: John Wiley & Sons, Inc., 2008.

Required Supplemental Reading:

Simpson, Pamela H. Cheap, *Quick, and Easy: Imitative Architectural Materials, 1870-1930*. Knoxville, TN: University of Tennessee Press; 1999.

Gayle, Margot, David W. Look, and John G. Waite. *Metals in America's Historic Buildings: Uses and Preservation Treatments.* Washington, D.C.: U.S. Department of the Interior, National Park Service, Cultural Resources, Preservation Assistance, 1992.

Additional readings must be ordered from the National Park Service and/or are available online. Ordering information will be provided by the instructor. Be sure to check the online calendar for links to online readings.

Basis for Student Evaluation:

Undergraduate Students:	Graduate Students
Midterm Exam 20%	Midterm Exam 15%
Final Exam 20%	Final Exam 15%
Building Condition Assessment 20%	Building Condition Assessment 20%
Diagnostic Observation Reports (2) 20%	Diagnostic Observation Reports (3) 30%
Architectural artifact analysis 10%	Architectural artifact analysis 10%
Class Participation, online and in-class Assignments and quizzes 10%	Class Participation, online and in-class assignments and quizzes 10%

Exams
Two "take home" exams, a midterm and a final, will be used to assess student mastery of the content covered in class readings and discussions.

Building Condition Assessment
Using the skills and methodologies discussed in class, students will individually or in teams prepare building condition assessment reports for a historic building located in the Cape Girardeau area. Details of the report will be discussed in class as the semester progresses, but the project is

intended to provide a practical experience for students to apply the knowledge and skills developed over the course of the semester.

Diagnostic Observation Reports
Drawing on the readings and class discussion, students will prepare diagnostic reports on deteriorated architectural features of their own choosing. Each report will focus on one element, preferably observable from a public place. Students will photograph the element, and write a diagnostic report explaining the probable cause (or causes) of the deterioration and recommended remediation steps.

Architectural artifact analysis
Students will choose an architectural artifact from the HP architectural artifact learning library in the HP dry lab and write a report detailing what the artifact is, when it was made, how it was used, what deterioration issues it has or could potentially have, and what remediation or care plan would be required if the artifact were to be returned to use.

Participation, online and in-class assignments and quizzes
Students will attend class regularly and complete various online and in-class short writing assignments and/or diagnostic exercises. Quizzes over the reading material may also be administered in class.

Note on Plagiarism: Not surprisingly, your paper (and all written assignments) must be your own work. This does not mean that you may not study together, share ideas and learn from one another. It simply means that you can not pass off the work of others as your own. Plagiarism and cheating are serious offenses and are discussed more fully in the University Bulletin. If you have any questions about what constitutes plagiarism or need clarification about any of the issues involved, please make an appointment to see me. This course abides by the university's policies regarding academic honesty.

Note: Assignments turned in late will be penalized 10%. In addition to participating in class discussion, students must complete all assignments to receive a grade for participation.

Course Policies and Procedures
This course abides by the university's policies regarding attendance, academic honesty, civility and harassment, and support for students with disabilities, and has additional requirements regarding the use of technology:

Attendance
• It is essential that you be in class. Class participation and discussion of the readings form an important part of the course. Students are expected to actively participate in class discussion, including commenting on ideas from the reading assignments, concepts presented by other students, and ideas advanced by the instructor. If you are not in class, you can't participate. In addition, we will have several guest speakers; experiences that cannot be duplicated if missed. The official statement about attendance, derived from the Undergraduate Bulletin, may be accessed at: http://www.semo.edu/bulletin/.

Academic Honesty
• Southeast Missouri State University expects all students, faculty and staff to operate in an honest and ethical manner. Academic dishonesty is a very serious offense because it undermines the value of your education and the education of others. Students who engage in academic dishonesty

face significant penalties. Forms of academic dishonesty include, but are not limited to, plagiarism, cheating, contract cheating, misrepresentation, and other actions you take. Some of these are defined below:

- o Plagiarism means passing off someone else's work as your own, whether it is intentional or unintentional.
- o Cheating includes copying from another person or source of information to meet the requirements of a task.
- o Contract cheating is paying someone else or a company to do your work.
- o Misrepresentation means you are posing as someone else or someone else is posing as you to complete a task.
- o Collusion means working with one or more people to cheat. If you help someone cheat or plagiarize you will face the same penalties.

For more information, visit the Responsible Redhawks Code of Conduct http://www.semo.edu/responsibleredhawks/code-of-conduct.html or the Faculty Handbook Section (D) on Academic Honesty http://www.semo.edu/facultysenate/handbook/5d.html

Accessibility

• Southeast Missouri State University and Disability Services are committed to making every reasonable educational accommodation for students who identify as people with disabilities. Many services and accommodations which aid a student's educational experience are available for students with various disabilities. Students are responsible for contacting Disability Services to register and access accommodations. Accommodations are implemented on a case by case basis. For more information, visit http://www.semo.edu/ds/ or contact Disability Services at 573-651-5927.

Civility

• Your university experience is purposely designed to introduce you to new ideas, help you think effectively, develop good communication skills, evaluate information successfully, distinguish among values and make sound judgements. Doing this well requires respectful and courteous discussion among and between students and the instructor. Together, we must create a space where we acknowledge and respect others have different experiences, perspectives and points of view. Disagreements are likely. Mutual respect for one another and a willingness to listen are important. Remember, you are responsible for your behavior and actions. There is a no tolerance policy on bullying or harassment of any kind. Additional information on student conduct may be found at: http://www.semo.edu/pdf/stuconduct-code-conduct.pdf?ver=1.0 and http://www.semo.edu/pdf/Conduct_Faculty_Resource_Guide.pdf

Mandatory Reporting

• I will keep information you share with me confidential to the best of my ability, but as a professor I am legally required to share information about sexual misconduct and crimes I learn about to make our campus and community safe for everyone.

Student Success

• This course uses SupportNET, Southeast's student success network, to improve communication between students, faculty, and staff on campus. You'll get emails through SupportNET with information about resources or concerns. Please read these emails—they are sent to help you succeed! You can access SupportNET through your portal, Moodle or directly at supportnet.semo.edu to see any academic alerts, ask for help and to access resources to support your success at Southeast.

Technology

• Students will be required to submit their papers electronically in the class Canvas page and complete several online assignments. Students also may be required to access some of the reading material for the course online. Students who anticipate difficulty in mastering any of the technological components of the course are responsible for seeking assistance from the instructor in a timely manner.

Cell Phone/Computer/Tablet Use in Class

• Students may use cell phones/computers/tablets during class only as part of a specific class assignment or to bring relevant information to class discussion. Any observed use of these devices without publicly sharing the information as part of class discussion will result in half credit for participation for that day. Research shows that "multitasking" interferes with learning, reduces comprehension and potentially lowers IQ. (see "Multitasking Damages Your Brain and Career, New Studies Suggest," http://www.forbes.com/sites/travisbradberry/2014/10/08/multitasking-damages-yourbrain-and-career-new-studies-suggest/#2e1391b22c16 (accessed 1/16/20).

• Students may use laptop computers to take notes but are discouraged from doing so. Studies show that taking notes on a laptop result in lower comprehension than taking notes long hand. (see "A Learning Secret: Don't Take Notes with a Laptop," https://www.scientificamerican.com/article/a-learning-secret-don-t-take-notes-with-alaptop/ (accessed 1/16/20).

Honors Program

• Are you in the Jane Stephens Honors Program and interested in creating an honors contract? I am an Honors faculty member and would be more than happy to work with you this semester. Please contact me the first two weeks of class through email, office hours, or set up an appointment.

University Policy Statement

• Questions, comments or requests regarding this course or program should be taken to your instructor. Unanswered questions or unresolved issues involving this class may be taken to Dr. Toni Alexander, Chair, Department of History and Anthropology.

COURSE OUTLINE AND SCHEDULE

Cohort attending in person	Date	Day	Topic
Group A	1/26	T	Course Introductions
Group B	1/28	Th	Part I: Process Overview Assignment: Young, Introduction and Chapter 1, pp. xi-16
Group A	2/2	T	Health and Life Safety Assignment: Young, Chapter 2, pp. 17-30 Preservation Briefs: 03: Improving Energy Efficiency in Historic Buildings.
Group B	2/4	Th	Building Pathology: Investigation, Analysis, and Assessment Assignment: Young, Chapter 3; Preservation Brief: 35: Understanding Old Buildings: The Process of Architectural Investigation, 39: Holding the Line: Controlling Unwanted Moisture in Historic Buildings.

Group A	2/9	T	Identifying Character-Defining Features, inside and out Assignment: Preservation Briefs 17: Architectural Character - Identifying the Visual Aspects of Historic Buildings as an Aid to Preserving Their Character, 18: Rehabilitating Interiors in Historic Buildings - Identifying Character-Defining Elements.
Group B and A	2/11	Th	Identifying Features and Conditions Fieldwork All students attend in person. We will meet at a pre-determined location.
Group A	2/16	T	Part II: Building Materials Wood Assignment: Young, Chapter 4, pp. 47-78; Preservation Brief: 26: The Preservation and Repair of Historic Log Buildings.
Group B	2/18	Th	Masonry Assignment: Young, Chapter 5, pp. 79-114; Preservation Briefs: 01: Assessing Cleaning and Water-Repellent Treatments for Historic Masonry Buildings, 02: Repointing Mortar Joints in Historic Masonry Buildings, 06: Dangers of Abrasive Cleaning to Historic Buildings, 07: The Preservation of Historic Glazed Architectural Terra-Cotta, 38: Removing Graffiti from Historic Masonry. Guest Instructor: Travis Ratermann, Survey Historian, Arkansas Heritage (SHPO) (and famous Southeast HP graduate).
Group A	2/23	T	No class: reading day
Group B	2/25	Th	Concrete Assignment: Young, Chapter 6, pp. 115-130; Preservation Briefs 15: Preservation of Historic Concrete, 42: The Maintenance, Repair and Replacement of Historic Cast Stone. Group A work on Diagnostic Observation Report
Group A	3/2	T	Concrete Assignment: Young, Chapter 6, pp. 115-130; Preservation Briefs 15: Preservation of Historic Concrete, 42: The Maintenance, Repair and Replacement of Historic Cast Stone. Group B work on Diagnostic Observation Report FIRST DIAGNOSTIC OBSERVATION REPORT DUE

Group B	3/4	Th	Architectural Metals Assignment: Young, Chapter 7, pp. 131-152; Preservation Brief 27: The Maintenance and Repair of Architectural Cast Iron
Group A	3/9	T	Architectural Metals II Assignment: Gayle, Waite and Look, Metals in America's Historic Buildings, Part I, pp. 5-88. Groups A & B Complete Online Canvas Assignment—Group A meets for metals workshop
Group B	3/11	Th	Architectural Metals III Assignment: Gayle, Waite and Look, Metals in America's Historic Buildings, Part II, pp. 89-153. Groups A & B Complete Online Canvas Assignment—Group B meets for metals workshop
Group A	3/16	T	Part III: Building Fabric Roofing Assignment: Young, Chapter 8, pp. 155-174; Preservation Briefs: 04: Roofing for Historic Buildings, 19: The Repair and Replacement of Historic Wooden Shingle Roofs, 29: The Repair, Replacement, and Maintenance of Historic Slate Roofs, 30: The Preservation and Repair of Historic Clay Tile Roofs
Group B	3/18	Th	Exterior Wall Cladding Assignment: Young, Chapter 9, pp. 175-198; Preservation Brief: 22: The Preservation and Repair of Historic Stucco. MIDTERM EXAM DUE
Group A	3/23	T	Imitative Architectural Materials Intro and Concrete Block Assignment: Simpson, Cheap, Quick and Easy, Intro and Chapter 1, pp. 1-29.
Group B	3/25	Th	Windows Assignment: Young, Chapter 10, pp. 199-220; Preservation Briefs: 09: The Repair of Historic Wooden Windows, 13: The Repair and Thermal Upgrading of Historic Steel Windows.

Group A	3/30	T	Metal Reading Assignment: Simpson, Cheap, Quick and Easy, Chapters 2 & 3, pp. 30-74. Groups A & B Complete Online Canvas Assignment—Group A meets for window workshop Presenter: Madelyn Brown (future famous Southeast HP graduate).
Group B	4/1	Th	Floors and Walls Assignment: Simpson, Cheap, Quick and Easy, Chapters 4 & 5, pp. 75-117. Groups A & B Complete Online Canvas Assignment— Group B meets for window workshop: Presenter: Madelyn Brown (future famous Southeast HP graduate).
Group A	4/6	T	Other Materials & Conclusions Assignment: Simpson, Cheap, Quick and Easy, Chapters 6 & 7, pp. 118-164.
Group B	4/8	Th	Cemetery Preservation Workshop Assignment: TBA Presenter: Madelyn Brown (future famous Southeast HP graduate). Field work on Saturday, April 10—headstone cleaning workshop
Group A	4/13	T	Entrances and Porches Assignment: Young, Chapter 11, pp. 221-238; Preservation Brief: 45: Preserving Historic Wooden Porches. SECOND DIAGNOSTIC OBSERVATION REPORT DUE
Group B	4/15	Th	Storefronts Assignment: Young, Chapter 12, pp. 239-152; Preservation Briefs: 11: Rehabilitating Historic Storefronts, 12: The Preservation of Historic Pigmented Structural Glass (Vitrolite and Carrara Glass).

Group A	4/20	T	Part IV: Building Ornamentals and Finishes Floors, Walls and Ceilings Assignment: Young, Chapter 13, pp. 255-270; Chapter 14, pp. 271- 28; an: Preservation Brief: 40: Preserving Historic Ceramic Tile Floors. All students attend in person for a concrete workshop. We will meet at a pre-determined location.
Group B	4/22	Th	Art and Stained Glass Assignment: Young, Chapter 15, pp. 289-302; Preservation Brief: 33: The Preservation and Repair of Historic Stained and Leaded Glass.
Group A	4/27	T	Wood Carving and Millwork Assignment: Young, Chapter 16, pp. 303-318; Preservation Brief: 34: Applied Decoration for Historic Interiors: Preserving Historic Composition Ornament. ARCHITECTURAL ARTIFACT ANALYSIS DUE
Group B	4/29	Th	Decorative and Flat Plaster Assignment: Young, Chapter 17, pp. 319-332; Preservation Briefs: 21: Repairing Historic Flat Plaster - Walls and Ceilings, 23: Preserving Historic Ornamental Plaster
Group A	5/4	T	Protective and Decorative Finishes Assignment: Young, Chapter 18, pp. 333-350; Preservation Briefs: 10: Exterior Paint Problems on Historic Woodwork, 28: Painting Historic Interiors.
Group B	5/6	Th	Putting it all together: Course Wrap-Up Assignment: TBA BUILDING ASSESSMENT DUE THIRD DIAGNOSTIC OBSERVATION REPORT DUE (Graduate Students)
	5/13	Th	FINAL EXAM (TAKE-HOME) DUE Final Exam due 12 pm (noon)

Course Syllabus
HP585-01: History of American Architecture
Southeast Missouri State University, Fall 2021
M, W, F 9:00-9:50, CR109

Dr. Steven J. Hoffman
Department of History
311 T Carnahan Hall
Office: (573) 651-2808
Cell: (573) 225-8393

Office Hours: T/Th 1:00-2:30
and by appointment
email: shoffman@semo.edu

www.facebook.com/hoffman.steven

Course Description

This course is a study of American architectural history as material evidence of the country's social, cultural, economic and technological development. (3 credit hours)

Purposes and Objectives of the Course

A. To introduce students to the history of various architectural styles and to be able to identify the significant characteristics of each style.
B. To create awareness of the relationship between architectural development and the social, cultural, economic, and technological forces of American life.
C. To enable students to relate the history of architecture and architectural styles to the broad patterns of national, regional and local growth.

Course Learning Outcomes:

A. Students will be able to identify the character-defining features of the major styles evidenced in the built environment of that portion of North America that presently comprises the United States of America.
B. Students will be able to explain ways in which the design and construction of vernacular architecture expresses individual and/or community identity.
C. Students will be able to explain the vernacular design process.

Expectations of Students

Students are expected to read all assignments, regularly attend class, participate in class discussions and/or activities, complete a written research paper, make one oral presentation and successfully complete all exams. Class participation and discussion of the readings form an important part of the course. Students are expected to participate actively in class discussion, including commenting on ideas from the reading assignments, concepts presented by other students, and ideas advanced by the instructor.

Basis for Student Evaluation

Students will be evaluated on class participation, a group project, three semester examinations, a cumulative final examination, a research paper, and one oral presentation.

Group Project

Students will divide into teams and conduct an architectural photo reconnaissance. Each team will present a poster panel of four or five architectural styles highlighting at least ten architectural features found in Cape Girardeau. Styles and features should be arranged (at least roughly) in some kind of chronological or developmental order, and should be organized to communicate a specific theme or idea about the historical patterns of development in Cape and/or environs. Presentations will take the form of a "poster session" which will be held during the last week of classes on Wednesday, December 8 (or Friday, December 10, depending upon when we schedule the fourth exam).

Historic Context Paper(s)
Each student will complete a 5-7 page historical context paper dealing with one of the styles or architects discussed in this course. NOTE: Graduate students will prepare two historical context papers, one on an architect AND one on an architectural style. Your paper MUST be on a recognized "high style" or prominent architect and MUST include references from the appropriate secondary literature—which means footnotes or in-text citations. Topics must be approved in advance.

Note: Encyclopedias, online and otherwise, are not strong sources for college papers. Likewise, you may use online architectural sources, such as www.greatbuildings.com, for some of your information, but these types of sites are also considered fairly weak sources—so be sure you don't rely on them too much for your paper. The text of your paper is to be 5-7 pages. If you illustrate your paper—which is encouraged—it will of course make the paper longer. A page of illustration is not equal to a page of text (just in case you were wondering) and will not count toward achieving the required page length.

Oral Presentations
Each student will complete a brief oral presentation over the course of the semester on a specific building from the canon of "great buildings." Students will choose from lists provided by the instructor. Presentations must use PowerPoint (or other presentation program) to present visual material pertaining to the subject, and should be 2 to 5 minutes in length. Dates of presentations will be determined by the instructor.

Note on Plagiarism: Not surprisingly, your paper (and all written assignments) must be your own work. This does not mean that you may not study together, share ideas and learn from one another. It simply means that you can not pass off the work of others as your own. Plagiarism and cheating are serious offenses and are discussed more fully in the University Bulletin. If you have any questions about what constitutes plagiarism or need clarification about any of the issues involved, please make an appointment to see me. This course abides by the university's policies regarding academic honesty. Penalties for academic dishonesty range from failing the assignment and/or failing the course to expulsion from the University.

The components of your grade will be weighted as follows:

Undergraduates:
Class Participation: 10%
Group Project 10%
First Exam: 10%
Second Exam: 10%
Third Exam 10%
Fourth Exam 10%
Final Exam 15%
Historical Context Paper: 20%
Oral Presentation: 5%

Graduate Students:
Class Participation: 10%
Group Project 10%
First Exam: 10%
Second Exam: 10%
Third Exam 10%
Fourth Exam 10%
Final Exam 15%
Historical Context Papers: 20%
Oral Presentation: 5%

Note: Assignments turned in late will be penalized 10%. In addition to participating in class discussion, students must complete all assignments to receive a grade for participation.

Course Policies and Procedures
This course abides by the university's policies regarding attendance, academic honesty, civility and harassment, and support for students with disabilities, and has additional requirements regarding the use of technology:

Attendance
It is essential that you be in class. Class participation and discussion of the readings form an important part of the course. Students are expected to actively participate in class discussion, including commenting on ideas from the reading assignments, concepts presented by other students, and ideas advanced by the instructor. If you are not in class, you can't participate. In addition, we will have several guest speakers; experiences that cannot be duplicated if missed. The official statement about attendance, derived from the Undergraduate Bulletin, may be accessed at: http://www.semo.edu/bulletin/.

Protect the Nest Classroom Guidelines [See previous syllabus for more on these areas]
Academic Honesty
Accessibility
Southeast Missouri State University and the Office of Accessibility Services are
Civility
Mandatory Reporting
Student Success
Technology
Cell Phone/Computer/Tablet Use in Class

Honors Program
Are you in the Jane Stephens Honors Program and interested in creating an honors contract? I am an Honors Faculty member and would be more than happy to work with you this semester. Please contact me the first two weeks of class through email, office hours, or set up an appointment.

University Policy Statement
Questions, comments or requests regarding this course or program should be taken to your instructor. Unanswered questions or unresolved issues involving this class may be taken to Dr. Toni Alexander, Chair, Department of History & Anthropology.

Textbooks
(Textbook available from Textbook Services; Additional Readings from Southeast Bookstore)

Gelernter, Mark. A History of American Architecture: Buildings in their Cultural and Technological Context. Hanover, NH and London, England: University Press of New England, 2001.

<u>Additional Reading</u> (available from Southeast Bookstore)

<u>Required</u>

Upton, Dell and Vlach, John Michael, eds. *Common Places: Readings in Vernacular Architecture*. Athens, GA: The University of Georgia Press, 1986.

<u>Recommended</u>

Blumenson, John J-G. *Identifying American Architecture: A Pictorial Guide to Styles and Terms, 1600-1945*. 3rd ed. New York: W.W. Norton & Company, Inc., 2012.

McAlester, Virginia and Lee. *A Field Guide to American Houses*. New York: Alfred A. Knopf; Expanded, Revised edition (November 10, 2015).
Burden, Ernest. *Illustrated Dictionary of Architecture*. 2nd Edition. New York: McGraw Hill, 2002.

Course Calendar:

Date	Day	Topic
Aug. 23	M	Course Introductions
Aug. 25	W	Before Europeans Assignment: Gelernter, Forward, pp. xviii-xxii and Chapter 1, pp. 1-34.
Aug. 27	F	European Colonial Architecture First European Arrivals: English Colonial: New England I Assignment: Gelernter, Chapter 2, pp. 35-42 and pp. 54-64.
Aug. 30	M	English Colonial: New England II Assignment: Norman Morrison Isham and Albert F. Brown, "Early Rhode Island Houses," and Robert Blair St. George, "'Set Thine House in Order": The Domestication of the Yeomanry in Seventeenth-Century New England," in Upton & Vlach, pp. 149- 157 and 336-366.
Sept. 1	W	English Colonial: The South Assignment: Fraser D. Neiman, "Domestic Architecture at the Clifts Plantation: The Social Context of Early Virginia Building," in Upton & Vlach, pp. 292-314; and, Dell Upton, "Vernacular Domestic Architecture in Eighteenth-Century Virginia," in Upton & Vlach, pp. 315-335.
Sept. 3	F	Readings in Vernacular Architecture I Assignment: "Introduction," and Fred B. Kniffen, "Folk Housing: Key to Diffusion," in Upton & Vlach, pp. xiii-26.
Sept. 6	M	Labor Day – No Class
Sept. 8	W	Readings in Vernacular Architecture II Assignment: Henry Glassie, "Eighteenth-Century Cultural Process in Delaware Valley Folk Building," and Thomas Hubka, "Just Folks Designing: Vernacular Designers and the Generation of Form," in Upton & Vlach, pp. 394-425 and 426-432.
Sept. 10	F	Georgian Architecture I Assignment: Gelernter, Chapter 3, pp. 65-85.
Sept. 13	M	Georgian Architecture II Assignment: Gelernter, Chapter 3, pp. 85-96.
Sept. 15	W	Spanish Colonial Assignment: Gelernter, Chapter 2, pp. 42-48

Sept. 17	F	French Colonial Assignment: Gelernter, Chapter 2, pp. 48-50.
Sept. 20	M	Dutch Colonial Assignment: Gelernter, Chapter 2, pp. 50-54.
Sept. 22	W	Readings in Vernacular Architecture III Assignment: Edward A. Chappell, "Acculturation in the Shenandoah Valley: Rhenish Houses of the Massanutten Settlement," in Upton & Vlach, pp. 27-57; and, Alan Gowans, "The Mansions of Alloways Creek," in Upton & Vlach, pp. 367-393.
Sept. 24	F	Preservation in Action—Internship Reports Assignment: TBA
Sept. 27	M	FIRST EXAM
Sept. 29	W	Searching for a National Style Federal Architecture Assignment: Gelernter, Chapter 4, pp. 97-115
Oct. 1	F	Readings in Vernacular Architecture IV Assignment: Fred B. Kniffen and Henry Glassie, "Building in Wood in the Eastern United States: A Time-Place Perspective," in Upton & Vlach, pp. 159-181; and, John Michael Vlach, "The Shotgun House: An African Architectural Legacy," in Upton & Vlach, pp. 58-78.
Oct. 4	M	Early Neoclassical Revival Assignment: Gelernter, Chapter 4, pp. 115-128.
Oct. 6	W	Greek Revival Assignment: Gelernter, Chapter 5, pp. 130-139.
Oct. 8	F	Fall Break -- No Class
Oct. 11	M	Mid-century revivalism Romantic Revivals I: Gothic Revival, Romanesque Revival, Italianate, Italian Villa, etc. Assignment: Gelernter, Chapter 5, pp. 139-165
Oct. 13	W	Romantic Revivals II: Renaissance Revival, Egyptian, etc. Assignment: Gelernter, Chapter 5, pp. 139-165.
Oct. 15	F	Romantic Revivals III
Oct. 18		Readings in Vernacular Architecture V Assignment: Catherine W. Bishir, "Jacob W. Holt: An American Builder, " in Upton & Vlach, pp. 447-481.
Oct. 20		SECOND EXAM
Oct. 22		Late-19th to early 20th century Victorian America I: Ruskin, High Victorian Gothic, Stick Style, Queen Anne, Eastlake, Shingle, Colonial Revival, Second Empire, Chateauesque, etc. Assignment: Gelernter, Chapter 6, pp. 166-189.
Oct. 25		Victorian America II Assignment: TBA

Oct. 27		Victorian America III Assignment: TBA
Oct. 29		H.H. Richardson
Nov. 1		American Renaissance
Nov. 3		Moving toward Modernism, and beyond Chicago School: First Skyscrapers Assignment: Gelernter, Chapter 7, pp. 190-229
Nov. 5		Frank Lloyd Wright
Nov. 8		THIRD EXAM
Nov. 10		Missouri German Vernacular
Nov. 12		HP Alumni Networking Workshop
Nov. 15		Toward the Modern: Art Deco & Streamline Moderne Assignment: Gelernter, Chapter 8, pp. 230-259.
Nov. 17		International Style and Modern I Assignment: Gelernter, Chapter 9, pp. 260-292.
Nov. 19		Modern II
Nov. 22		Commercial Architecture on Main Street
Nov. 24		Thanksgiving Holiday -- No Class
Nov. 26		Thanksgiving Holiday -- No Class
Nov. 29		Late Modern/Post-Modern Assignment: Gelernter, Chapter 10, 293-318.
Dec. 1		Post-Modern
Dec. 3		Post World War II Housing Styles
Dec. 6		Post World War II Housing Styles Assignment: TBA DUE: Research paper due in class
Dec. 8		Group Project Presentations
Dec. 10		FOURTH EXAM
Dec. 15		FINAL EXAM (Comprehensive Style Exam) Note: Exam Time is 8 am

Course Syllabus
HP588: Legal and Economic Principals of Historic Preservation
Southeast Missouri State University, Spring 2021
T/Th 9:30-10:45, CR109

Dr. Steven J. Hoffman
Department of History
311 T Carnahan Hall
Office: (573) 651-2808
Cell: (573) 225-8393

Office Hours: T/Th 1:00-2:30
and by appointment
email: shoffman@semo.edu

www.facebook.com/hoffman.steven

Catalog Description and Credit Hours of Course:
A study of the legal and economic aspects in the procedures for locating, recording and preserving America's historical, architectural and cultural resources. (3 hours)

Prerequisite(s):HP100, HP200, HP585

Purposes or Objectives of the Course:
A. To familiarize students with the development of historic preservation legislation in the United States.
B. To develop a foundation in the primary legal and economic issues facing historic preservation in the United States.
C. To understand the basic legal concepts and procedures implemented by federal, state, and local governments to locate, record, restore, and preserve the historic built environment.
D. To gain practical experience with the National Register of Historic Places and in conducting Historic Resource Survey work.

Student Learning Outcomes:
A. Students will be able to describe the role of federal, state, and local government in historic preservation.
B. Students will be able to prepare successfully a National Register of Historic Places nomination.
C. Students will be able to prepare successfully a Historic Resource Survey.

Expectation of Students:
Students are expected to attend all classes, do all required readings, demonstrate an acceptable degree of understanding on written examinations and/or papers, and participate in individual, group and class projects.

COVID-19 Updates:
As part of our COVID-19 accommodations we will be dividing the class into two cohorts: We will have a Tuesday cohort (Group A) and a Thursday cohort (Group B). The two cohorts will alternate class and remote (synchronous) attendance.
Remote participation: Students will be expected to participate remotely using Zoom during regularly scheduled class times. Students should, to the extent feasible, keep their camera on during class to help foster a sense of class community. Please let the professor know in advance if there is a reason for not having the camera on during any particular class session. Class sessions will be recorded and posted in theOffice365 Teams/Stream folder for the class.

Undergraduates:		Graduate Students:	
Midterm Exam:	10%	Midterm Exam:	10%
Final Exam	15%	Final Exam	15%
Survey Project	15%	Survey Project	15%
Most Endangered Building Listing	10%	Most Endangered Building Listing	10%
National Register Nomination	25%	National Register Nomination	25%
Online and in class assignments, drafts and quizzes	5%	Online and in class assignments, drafts and quizzes	5%
Class Participation	10%	Class Participation	10%

*Critical Response Paper: Graduate students must EITHER complete a 5-7 page critical response paper examining a legal or economic topic in historic preservation (e.g., important court case, tax credits, etc.), OR, complete a 5-7 page context paper that could be used as part of a commercial Historic Resource Surveyor National Register District or Multiple Property Submission in Cape Girardeau. The critical response (or context) paper is due by the date of the final exam.

Note: Included as part of the class participation grade, all students (undergraduate and graduate) must attend at least one session of the Cape Girardeau Historic Preservation Commission and submit a one-page response paper outlining the issues discussed at the meeting as well as your own observations of the process .The Commission meets on the third Wednesday of each month in the City Council Chambers. Students who cannot attend one of these meetings MUST make alternate arrangements with the instructor in a timely fashion to attend the meeting of a similar body.

Note: Assignments turned in late will be penalized 10%.In addition to participating in class discussion, students must complete all assignments to receive a grade for participation.

Protect the Nest Classroom Guidelines [See previous syllabus for more on these areas]
Academic Honesty
Accessibility
Southeast Missouri State University and the Office of Accessibility Services are
Civility
Mandatory Reporting
Student Success
Technology
Cell Phone/Computer/Tablet Use in Class

Textbook(s) and/or Other Required Materials or Equipment:

Textbook Rental:

Norman Tyler, *Historic Preservation: An Introduction to its History, Principles, and Practice*

Required Supplemental Reading:

Thomas King, *Cultural Resource Laws and Practice: An Introductory Guide*
Donovan Rypkema, *The Economics of Historic Preservation* (National Trust for Historic Preservation)**

Christopher J. Duerksen and Richard J. Roddewig, *Takings Law in Plain English*, (National Trust for Historic Preservation)***

Jane F. Boyle, Stuart Ginsberg and Sally G. Oldham, revised by Donovan D. Rypkema, *A Guide to Tax-Advantaged Rehabilitation* (National Trust for Historic Preservation).***

**this text is available through Amazon.
***these texts are temporarily unavailable for purchase. You can access a pdf of the readings on Canvas.

Additional readings are available on-line from the National Park Service.

COURSE OUTLINE AND SCHEDULE

Cohort attending in person	**Date**	**Day**	**Topic**
Group A	1/26	T	Course Introductions
Group B	1/28	Th	Background for American Preservation Legislation Assignment: Tyler, Historic Preservation, Forward, Acknowledgements, Chapters 1, 2 and 3, pp. ix-67; and King, Chapter 1, pp. 1-54.
Group A	2/2	T	The National Register of Historic Places Assignment: Tyler, Historic Preservation, Chapter 5, pp. 95-125; and King, Chapter 3, pp. 87-107.
Group B	2/4	Th	The National Register of Historic Places Assignment: National Register Bulletin No. 15, "How to Apply the National Register Criteria for Evaluation".
Group A	2/9	T	National Register (con't) Assignment: National Register Bulletin No. 16A, "How to Complete the National Register Registration Form". Also quickly review National Register Bulletin No. 16B, "How to Complete the National Register Multiple Property Documentation Form".
Group B	2/11	Th	National Register (con't) Assignment: Review National Register Bulletin No. 16A, "How to Complete the National Register Registration Form," paying particular attention to completing sections 7 and 8; National Register Bulletin No. 21, "Defining Boundaries for National Register Properties".
Group A	2/16	T	National Register projects Assignment: National Register Bulletin No. 21, "Defining Boundaries for National Register Properties". Background Reading: Review National Register of Historic Places Program: Sample Nominations (http://www.nps.gov/nr/sample_nominations.htm)In Class: 2 Minute presentations on National Register Property and Preliminary Statement of Significance

Group B	2/18	Th	Reading, Researching and Photographing a Building for the National Register Assignment: Tyler, Historic Preservation, Chapter 8 (partial), pp-214-231; National Register Bulletin No. 23, "How to Improve the Quality of Photos for National Register Nominations"; HABS Standards for Photographs and NPS Digital Photograph Policy; National Register Bulletin No. 39, "Researching a Historic Property".(https://www.nps.gov/subjects/nationalregister/publications.htm)Review: Barbara J. Howe, Delores A. Fleming, Emory L. Kemp, and Ruth Ann Overbeck, Houses and Homes: Exploring Their History(Nashville, TN: American Association of State and Local History, 1987), pp. 1-158 (this is the text you used in HP200). Guest Instructor: Travis Ratermann, Survey Historian, Arkansas Heritage (SHPO) (and famous Southeast HP program graduate). (and famous Southeast HP graduate).
Group A	2/23	T	Reading Day –No Class
Group B	2/25	Th	State Preservation Law Assignment: Review the website for the National Conference of State Historic Preservation Officers (available online http://www.ncshpo.org/about/).Complete assignment online (Canvas): Class does not meet today
Group A	3/2	T	Conducting a Local Survey: Standards and Techniques Assignment: National Register Bulletin No. 24, "Guidelines for Local Surveys: A Basis for Preservation Planning".
Group B	3/4	Th	Survey Workshop: Good Forms, Bad Forms and Tax Forms Assignment: Choose a house (any house) and complete a survey form for it. Bring the completed survey form to class. NOTE: Endangered Building nomination due March 5, 2021
Group A	3/9	T	Review and Compliance, NHPA, NEPA, Advisory Council Assignment: "Protecting Historic Properties: A Citizen's Guide to Section 106 Review" (download from Canvas) and King, Chapter 2, pp. 55-85, and Chapter 4, pp. 105-208.
Group B	3/11	Th	Review and Compliance (con't) Assignment: King, Chapter 5 [partial—on section 4(f)] pp. 240-244. Guest Speaker: Ms. Karen Daniels, Architectural Historian, Missouri Department of Transportation (and famous Southeast HP program graduate).

Group A	3/16	T	National Register Workshop: Presentation and Critique of National Register properties. Guest Speaker: April Scott, National Register and Survey Specialist, State Historic Preservation Office, Missouri Department of Natural Resources (and famous Southeast HP program graduate). DRAFT OF SECTION 7, ARCHITECTURAL DESCRIPTION OF NATIONAL REGISTER NOMINATION, DUE IN CLASS
Group B	3/18	Th	Preservation Planning Assignment: Tyler, Historic Preservation, Chapter 9, pp. 255-267; and Bradford J. White and Richard J. Roddewig, Preparing a Historic Preservation Plan(American Planning Association, Planning Advisory Service Report No. 450, 1994), pp. 1-20 (excerpt available from online syllabus), and review City of Cape Girardeau Historic Preservation Plan (available online at: http://www.cityofcapegirardeau.org/CityHall-PLANNING/Historic-Preservation-Master-Plan.aspx). Complete assignment online (Canvas): Class does not meet today MIDTERM EXAM (TAKE-HOME) UPLOAD TO CANVAS
Group A	3/23	T	Legal Basis for Historic Preservation Assignment: Tyler, Historic Preservation, Chapter 4, pp. 69-93; and Stephen A. Morris, "Zoning and Historic Preservation" (available online at nps.gov/history/hps/pad/partnership/Zoning699.pdf). Optional reading: Julia H. Miller, "A Layperson's Guide to Historic Preservation Law."
Group B	3/25	Th	Preservation at the Local Level: Where the Action Is! Assignment: Tyler, Historic Preservation, Chapter 6, pp. 127-159; Robert E. Stipe, "A Letter to George, or How to Keep the Preservation Commission Out of Court and Avoid Being Sued"; and City of Cape Girardeau Historic Preservation Ordinancehttp://www.municode.com/resources/gateway.asp?pid=10505&sid=25). Note: You will need to scroll down to "chapter 30 Zoning" and then choose "Article 2 Administration, Enforcement and Review," choose "Sec. 30-203 Historic preservation commission." To see Historic District regulations, under chapter 30 Zoning choose "District Regulations" then choose "Sec. 30-352 historic overlay district." Guest Speaker: Ryan Shrimplin, City Planner, City of Cape Girardeau, MO. DRAFT OF NATIONAL REGISTER FORM JACKET, UPLOAD TO CANVAS

Group A	3/30	T	Survey Workshop: Presentation of Findings Assignment: Survey Teams will present three (3) representative houses/buildings to the class (houses should be as different from each other as possible in the attempt to get full range of represented styles). DRAFT SURVEY FORMS (WITH PHOTOS) TO SURVEY COORDINATORS
Group B	4/1	Th	Survey Workshop: Presentation and critique (con't) Assignment: Review and comment on draft survey forms.
Group A	4/6	T	Preservation in Action –Main Street Assignment: Tyler, Historic Preservation, Chapter 9 (partial), pp. 268-275. Guest Speaker: Katelyn Brotherton, Resource Development Coordinator, Missouri Main Street Connection (and famous Southeast HP graduate), and/or Ben White, Program OutreachSpecialist, Missouri Main Street Connection, Former Executive Director, Chillicothe Main Street, Chillicothe, Missouri (and famous Southeast HP graduate).
Group B	4/8	Th	Preservation in Action –Evaluating and Describing Historic Resources Assignment: TBA Guest Speaker: Laura Williams, Historic Resource Specialist, Illinois State Archeological Survey (and famous Southeast HP graduate).
Group A	4/13	T	Reports from the Real World: Public Archeology Guest Speaker: Heather Carey, Archeologist, Shawnee National Forest, USDA National Forest Service (and famous Southeast HP graduate). DRAFT OF SECTION 8, STATEMENT OF SIGNIFICANCE FOR NATIONAL REGISTER NOMINATION, UPLOAD TO CANVAS

Group B	4/15	Th	NAGPRA Assignment: King, Chapter 6, pp. 251-286. Guest Speaker: Lara Anderson and Heather Pobst (famous Southeast HP graduate), U.S. Army Corps. Of Engineers, St. Louis Division.
Group A	4/20	T	Tax Incentives for Historic Preservation –History and Current Status Assignment: Tyler, Historic Preservation, Chapter 10, pp. 301-317; Boyle, Ginsberg and Oldham, "A Guide to Tax-Advantaged Rehabilitation," and, NPS Historic Preservation Tax Credits (available online: http://www.nps.gov/tps/tax-incentives.htm). Guest Speaker: Ms. Riley Price, Interim Executive Director, Missouri Preservation (and famous Southeast HP graduate).
Group B	4/22	Th	COMPREHENSIVE SURVEY PROJECT DUE
Group A	4/27	T	The Economics of Historic Preservation Assignment: Donovan Rypkema, Economics of Preservation, pp. 1-112.
Group B	4/29	Th	The Takings Issue Assignment: Christopher J. Duerksen and Richard J. Roddewig, "Takings Law in Plain English" (National Trust for Historic Preservation).
Group A	5/4	T	Design Review Assignment: Tyler, Historic Preservation, Chapter 9 (partial), pp. 275-288, and "Working on the Past in Local Historic Districts" (http://www.nps.gov/tps/education/workingonthepast/index.htm) "Section C: Creating and Using Design Guidelines" (the role they play, accepted principles, etc.). You can view C at http://www.nps.gov/tps/education/workingonthepast/roletheyplay.htm, or if you prefer you can print out the entire section C at http://www.nps.gov/tps/education/workingonthepast/sectionc.pdf). Also review Raleigh's design guidelines (you can find it under the section in the box entitled: EXAMPLE OF AN ILLUSTRATED LOCAL DESIGN GUIDELINE or go directly to http://www.nps.gov/tps/education/workingonthepast/raleigh_guidelines.pdf). Section C is part of NPS' larger Working on the Past in Local Historic Districts, a nice resource in its own right (available online). See also the National Alliance of Preservation Commissions online design guidelines at http://napc.uga.edu/resources-links/online-design-guidelines/for additional examples
Group B	5/6	Th	Where to From Here? Assignment: King, Chapter 9, pp. 343-364. NATIONAL REGISTER NOMINATION DUE
Group A and B	5/13	T	FINAL EXAM (TAKE-HOME) DUE Final Exam due 10 am

Course Syllabus
HP589: Historic Preservation-Based Economic Revitalization
Southeast Missouri State University, Spring 2020
T/Th 11:00-12:15, CR109

Dr. Steven J. Hoffman
Department of History
311 T Carnahan Hall
Office: (573) 651-2808
Cell: (573) 225-8393

Office Hours: T/Th 12:30-2:00
and by appointment
email: shoffman@semo.edu

www.facebook.com/hoffman.steven

Catalog Description and Credit Hours of Course:

A study of the role of historic preservation in the revitalization of America's historic downtowns. (3 hours)

Purposes or Objectives of the Course:

In this course students will:

A. Explore the history of downtown commercial districts and the various factors that account for their development during the nineteenth and early twentieth centuries and their eventual decline after World War II.
B. Be able to identify the typical building types and development patterns that constitute traditional downtown commercial districts.
C. Analyze and discuss the various strategies used to promote downtown revitalization in response to the changes occurring after World War II.
D. Demonstrate a working knowledge of the principles of the historic preservation-based strategy for downtown revitalization embodied in the National Trust for Historic Preservation's Four Point Main Street Approach.
E. Demonstrate the ability to apply the theory of historic preservation-based economic revitalization to the problems found in the cities and towns of contemporary America
F. Gain practical experience working with the concepts of the Main Street Four Point Approach.

Student Learning Outcomes:

A. Students will be able to identify the typical building types and development patterns that constitute traditional downtown commercial districts.
B. Students will demonstrate a working knowledge of the principles of the historic preservation-based strategy for downtown revitalization embodied in the National Trust for Historic Preservation's Four Point Main Street approach.
C. Students will demonstrate familiarity with the required elements necessary for successful completion of a feasibility assessment for rehabilitation of vacant historic buildings.

Expectation of Students:

A. Attend class regularly.
B. Actively participate in class discussion.
C. Read all materials assigned.
D. Complete all written and oral assignments in a timely manner.
E. Demonstrate mastery of course content on examinations and/or other assessments.
F. Complete a comprehensive course project.

Basis for Student Evaluation:

Undergraduate Students:		*Graduate Students*	
Midterm Exam	15%	Midterm Exam	10%
Final Exam	15%	Final Exam	10%
Team Project*	20%	Team Project*	20%
Feasibility Assessment**	30%	Feasibility Assessment**	30%
Class Participation***	20%	Class Participation***	20%
Analytical Paper****	10%*		

***Team Project:** The class as a whole will work with Old Town Cape to develop and present a walking tour event. Details about the project will be presented in class on January 30.

****Feasibility Assessment**: Students will work in teams to prepare a feasibility assessment for a historic property using Donovan Rypkema's Feasibility Assessment Manual for Reusing Historic Buildingsas a guide.

*****Participation:** Students are expected to participate not only in class discussion (and demonstrate command of the material), but to engage actively during all field sessions, utilizing them fullyas learning opportunities, participate fully and completely in the group project, and complete any assignments or out-of-class activities assigned during the semester.

******Analytical Paper**: Graduate students must complete a 5-7 page analytical paper examining the needs of a Main Street community and offering recommendations for committee action plans. The paper is due by the date of the final exam.

Directions for Graduate Paper:

1.Interview the Main Street Manager or Board President (or both) of a local Main Street program to determine what they consider to be:
   a. the most significant accomplishment of the program in the last yearb.
   b. what the top three areas or projects that need to be addressed by the organization in the coming year
   c. the likelihood of actually accomplishing those goals.
2.Conduct a field survey of the district (or portion of the district) to identify what you believe the strengths, weaknesses, opportunities and threats are.
3.Write a 5-7 page paper outlining your findings and suggesting one work plan project for each of the four committees that you believe would help advance the revitalization effort.

Note: Assignments turned in late will be penalized 10%.In addition to participating in class discussion, students must complete all assignments to receive a grade for participation.

Note on Plagiarism:[See previous syllabus for more on this.]

Protect the Nest Classroom Guidelines [See previous syllabus for more on these areas]

Academic Honesty

Accessibility

Southeast Missouri State University and the Office of Accessibility Services are

Civility

Mandatory Reporting

Student Success

Technology

Cell Phone/Computer/Tablet Use in Class

Textbook(s) and/or Other Required Materials or Equipment:

PDF in Moodle:

Dono, Andrea L. *Revitalizing Main Street: A Practitioner's Guide to Comprehensive Commercial District Revitalization.* Washington, DC: National Trust Main Street Center,2009.

Required Supplemental Reading:

Francaviglia, Richard. *Main Street Revisited: Time, Space, and Image Building in Small-Town America.* Iowa City: University of Iowa Press, 1996.

Rypkema, Donovan. *Feasibility Assessment Manual for Reusing Historic Buildings.* CreateSpace Independent Publishing Platform (December 29, 2015).

Various additional readings made available through the Moodle page.

COURSE OUTLINE AND SCHEDULE

Date	Day	Topic
1/21	T	Course Introductions
1/23	Th	Main Street America: 4 points Assignment: Main Street America, "The Main Street Approach: A Comprehensive Guide to Community Revitalization"
1/28	T	Main Street America: picking transformation strategies Assignment: TBA
1/30	Th	Old Town Cape Historic Preservation Month Project Assignment: TBA Guest Speaker: Sarah LaVenture, Assistant Director, Old Town Cape (and famous Southeast HP graduate); Andrea Hamm, Old Town Cape student worker (and future famous Southeast HP graduate)
2/4	T	Feasibility Assessments Assignment: Donovan Rypkema, Feasibility Assessment Manual for Reusing Historic Buildings, pp. 1-89 (entire book).
2/6	Th	Feasibility Assessments Assignment: Donovan Rypkema, Feasibility Assessment Manual for Reusing Historic Buildings, pp. 1-89 (entire book).
2/11	T	Project Building Tour: 101 William Street, Cape Girardeau, MO
2/13	Th	Understanding Downtown Assignment: Main Street Revisited, Foreword, Introduction, and Chapter 1, pp. xi-64.
2/18	T	Understanding Downtown Assignment: Main Street Revisited, Chapter 2, pp. 65-129.

2/20	Th	Downtown District Analysis Assignment: Field Analysis
2/25	T	Analysis debrief
2/27	Th	Organization: Action Plan Workshop Guest Speaker: Bridgette Kelch, Director of Development, East Central College, Executive Director, East Central College Foundation AND Former Executive Director, Downtown Washington, Washington, Missouri's Main Street program (and famous Southeast HP graduate).
3/3	T	Using Market and Demographic Data Assignment: TBA Guest Speaker: Keith Winge, State Community Development Director, Missouri Main Street Connection
3/5	Th	Understanding Downtown Assignment: Main Street Revisited, Chapter 3, pp. 130-192.
3/10	T	Understanding Downtown Downtown Walking Tour
3/12	Th	Team Project Planning time Assignment: Meet as needed for OTC or Feasibility Team projects Midterm Exam Due (Take Home)
3/17	T	Spring Break
3/19	Th	Spring Break
3/24	T	History Day: No class—plan to volunteer as a History Day judge instead.
3/26	Th	Economic Vitality: Real Estate Development from the Developer's Perspective Assignment: Revitalizing Main Street, Chapter 10, pp. 93-102. Guest Speaker: Jeff Mauer, President, Mayson Capital Partners DUE: Feasibility Assessment Project ideas to design teams
3/31	T	Promotion: Building Excitement Assignment: Revitalizing Main Street, Chapter 21, pp. 195-212. Guest Speaker: Ben White, Program Outreach Specialist, Missouri Main Street Connection program (and famous Southeast HP graduate).
4/2	Th	Feasibility Assessment Workshop: Financing Development Projects; A Banker's Perspective Guest Speaker: Blake Lingle, Commercial Loan Officer, The Bank of Missouri
4/7	T	Working with/for Boards and Engaging Volunteers Assignment: Revitalizing Main Street, Chapter 4, pp. 25-36, and "Volunteer Management: Meeting the Challenge and Raising the Bar."

4/9	Th	Entrepreneurial Ecosystems Assignment: Main Street America, "Entrepreneurial Ecosystems and the Role of Commercial Districts"
4/14	T	Tour of Marquette Tech District Guest Speaker: Dr. James Stapleton, Co-founder, Codefi, Marquette Tech District
4/16	Th	Community Engagement for Community Transformation Assignment: MSA, Community Engagement for Community Transformation
4/21	T	The State of Main Street in Missouri and the Nation Guest Speaker: Gayla Roten, Executive Director, Missouri Main Street Connection Assignment: TBA
4/23	Th	Leadership Development Workshop Assignment: TBA Draft of Feasibility Assessment Due
4/28	T	Feasibility Assessment workshop Assignment: Discuss draft feasibility assessments
4/30	Th	The Future of Retail Assignment: MSA, The Future of Retail: Creative Approaches to Place-Based Entrepreneurship Guest Speaker: Laurie Everett, owner Annie Laurie's, Indie House, etc.
5/5	T	Project Presentations Present Feasibility Assessments to client Final Feasibility Assessment Due
5/7	Th	Final Projects Work Day Team Projects Due
5/14	Th	FINAL EXAM (TAKE-HOME) DUE Graduate Student Analytical Papers Due Exam due 12:00 pm (noon)

6

Program Growth and Expansion

The 17th edition of *Pastscript*, 1991-1992, reported that the SEMO HP Program continued to grow, with 31 new students from all over the U.S. entering the program in fall 1991. Among them were three National Merit Finalists.

In Spring 1993, there were 87 historic preservation majors enrolled, with 31 entering the undergraduate program earlier that fall (1992), according to the 18th edition of *Pastscript*, 1992-1993.

Among the changes made by the Department of History in 1992-1993, "was a revision to the graduate program, adding an area of emphasis in Historic Preservation with a Heritage Education focus. The other two areas of emphasis, Research and Teaching, were slightly revamped to assure that student expectations in all areas of emphasis are similar. To our knowledge, we are the first institution in the country to develop a program emphasizing Heritage Education, a newly emerging area of study." "In addition to the current program of courses, the [History] Department's new historic preservation emphasis for the Master of Arts degree should be in place for the Fall 1993 semester."[57]

The 2007 *Pastscript* reported 60 majors.

Graduate / Master's Degree Added

Southeast Missouri State University offers both undergraduate and graduate degree programs in historic preservation. Undergraduates may earn the bachelor of science degree in historic preservation.

By August 1996, Historic Preservation was "offered as a minor for students majoring in anthropology, interior design, history, or other fields. Historic Preservation majors may now choose to minor in anthropology, archaeology, architectural design, history, interior design, or promotion management."[58]

Graduate students today (2021) have options.

Dr. Steven Hoffman shared some institutional knowledge with which he has a personal connection from his tenure at SEMO.

"The Master of Arts in History degree was modified in the early 1990s to allow for students to earn the degree with an emphasis in Historic Preservation. In 2010, Drs. Steven Hoffman and Daryl Fridley collaborated to create a new Master of Arts degree in Public History with emphases in Historic Preservation and Heritage Education."

The Public History Graduate Program created and debuted their Facebook page on August 30, 2011.

Hoffman added, "The department continued to offer a separate MA in History degree until it was discontinued in 2020.

"The Heritage Education emphasis was modified in 2016 to Heritage Interpretation after Dr. Fridley left the department. It changed again in 2020 to an emphasis in Digital Heritage.

"[Dr. Hoffman] developed the 4+1 accelerated MA program in 2016 allowing undergraduate students to complete both their BS and MA in Public History degrees in five years, graduating the first students with the accelerated degree in 2021."[59]

v1993 graduates David W. Jackson (left) and Tony Ford (right)
>Jackson with his maternal Great Grandfather and Grandparents

7

Historic Preservation Association Banquets and Awards

"The Historic Preservation Association (HPA) strives to bring all students together interested in history and the preservation of the past. We work to provide fun and educational experiences through trips to historic sites, lectures from those in the field, and various other volunteer activities."[60]

Today (2021), SEMO directs browsers to the HPA's public Facebook group page that was created on February 15, 2009. There are 322 members of this page as of September 2021 (facebook.com/groups/127604600314/).

The following was transcribed from the "Way Back Machine," at archive.org from "way back" in 2005-2006.[61]

Welcome to our website!

The Historic Preservation Association is a student group here at Southeast Missouri State University comprised of undergraduate and graduate students in historic preservation, history, anthropology and other related (and unrelated) disciplines.

Our mission is to:

- bring together beginning and advanced students of historic preservation with faculty and community leaders;
- encourage the spirit of regional preservation of historic sites, artifacts, and cultures that will reflect ongoing trends nationwide;
- and to foster involvement with preservation activities and various other organizations while providing resources for higher education and job opportunities.

We plan volunteer projects and field trips to regional sites such as Ste. Genevieve, Missouri, and Wickliffe Mounds, Kentucky, so our members have a chance to interact and experience the area, people, and profession.

We also hold an annual Banquet to wrap up our school year and reflect on all we have done! A guest speaker is invited to give a short talk on their involvement in historic preservation and offer a personal view of projects and experiences.

SEMO HP Alumni at the HPA Banquet, 2006

We encourage anyone with an interest in history, preservation, or getting involved with a great group of individuals to stop by one of our meetings or email us and let us know about yourself!

Please make sure you read through the latest meeting minutes and check out our updated pictures of the events, people, and projects.

Thank you so much for visiting this website!

2005-2006 Officers

President: Robyn Mainor
Vice-President: Travis Ratermann
Secretary: Diane Riley
Treasurer: Allison Marshaus
Historian: Catie Myers
Graduate Representative: Bill Gatlin
Faculty Advisor: Dr. Eric Clements (pictured below)

> Dr. Eric Clements
HPA Faculty Advisor since 1999

The following compilation of HPA officers is possible largely through the efforts of Allyson Ingli, HPA Vice President in 2020-2021 and 2021-2022, with much appreciation and gratitude from the author.

If you have ANYTHING original from your time at SEMO that has any remote connection to the HP Program and/or HPA, please consider donating ORIGINALS for posterity...you can keep copies/scans for yourself! Start at semohpalumni.org.

SEMO HPA OFFICERS, 1982-2022

If you have some *documentation* to add to or correct the following, go to semohpalumni.org.

1982-1983
President: Debbie Reid
Secretary: Lisa Couch
Historian: Melanie Gimpel
Faculty Advisor: Dr. Art Mattingly

SEMO HP 1st Anniversary
Vice-President: Mary Malone
Treasurer: Karen Luker

1983-1984
President:
Secretary:
Historian
Faculty Advisor: Dr. Art Mattingly

Vice-President: Bob Ulrich
Treasurer:

1984-1985
President: Claudia Finley
Secretary: Jeanette Hultberg
Historian: Frank Woodson
Faculty Advisor: Dr. Art Mattingly

Vice-President: Renae Farris
Treasurer: Joan Feezor
Parliamentarian: Scott Raines

1985-1986
President: Frank Strassler
Secretary:
Historian:
Faculty Advisor: Dr. Art Mattingly

Vice-President: Creager Smith
Treasurer:
Parlimentarian:

1986-1987
President: Creager Smith
Secretary: Joan Feezor
Faculty Advisor: Dr. Art Mattingly

SEMO HP 5th Anniversary
Vice-President: Tom Neff
Treasurer: Frank Strassler
Hosted the first HPA banquet, per Creager Smith

1987-1988
President: Charles Pautler
Secretary: Karen Daniels
Faculty Advisor: Dr. Art Mattingly

Vice-President: Jennifer McNeely
Treasurer: Creager Smith

1988-1989
President: Susan Haake
Secretary/Treasurer: Mike Dalman
Faculty Advisor: Dr. Art Mattingly

Vice-President: Melissa Dirr

1989-1990
President:
Secretary:
Historian:
Faculty Advisor:

Vice-President:
Treasurer:

1990-1991
President: Brian Keith McCutchen
Secretary:
Historian: Felix Kinsley
Faculty Advisor:
Vice-President: Sue Corvick
Treasurer:

1991-1992
President:
Secretary:
Historian:
Faculty Advisor:
SEMO HP 10th Anniversary
Vice-President:
Treasurer:

1992-1993
President:
Secretary:
Historian:
Faculty Advisor:
Vice-President:
Treasurer:

1993-1994
President:
Secretary:
Historian:
Faculty Advisor:
Vice-President:
Treasurer:

1994-1995
President: Sandra Shupp
Secretary: Becky Rolwes
Historian: ---------------
Faculty Advisor:
Vice-President: Mike Highfield
Treasurer: Beth Pate

1995-1996
President: Valerie Gaumont
Secretary: Anne Kern
Historian: -----------
Faculty Advisor:
Vice-President: Scott Richard "Scottie" Cross
Treasurer: Beth Pate
Parlimentarian: Sandra Shupp

1996-1997
President:
Secretary:
Historian:
Faculty Advisor:
SEMO HP 15th Anniversary
Vice-President:
Treasurer:
Graduate Representative:

1997-1998
President: Kim Cleveland
Secretary: Mandy Wagoner
Historian: ------
Archivist: Amy Derscheid
Faculty Advisor:
Vice-President: Lee Rains
Treasurer: Stacey Cummins
Graduate Representative: Renae Farris
Parlimentarian: Gretchen Greminger

1998-1999
President:
Vice-President:
Secretary:
Treasurer:
Historian:
Graduate Representative:
Faculty Advisor:

1999-2000
President: Donna Rausch
Vice-President: Rich Jackson
Secretary:
Treasurer: Becky Garnaat
Historian:
Graduate Representative: Landon Smith
Faculty Advisor: Dr. Eric Cements
Parlimentarian: Matt Andres

2000-2001
President:
Vice-President:
Secretary:
Treasurer:
Historian:
Graduate Representative:
Faculty Advisor: Dr. Eric Cements

2001-2002
SEMO HP 20th Anniversary
President:
Vice-President:
Secretary:
Treasurer:
Historian:
Graduate Representative:
Faculty Advisor: Dr. Eric Cements

2002-2003
President: Bob Bettis
Vice-President:
Secretary:
Treasurer:
Historian:
Graduate Representative:
Faculty Advisor: Dr. Eric Clements

2003-2004
President:
Vice-President:
Secretary:
Treasurer:
Historian:
Graduate Representative:
Faculty Advisor: Dr. Eric Cements

2004-2005
President: Robyn Mainor
Vice-President: Garrett Wright
Secretary: Diane Sanders
Treasurer: Allison Ann
Historian: Catie Myers
Graduate Representative: Lisa Graham
Faculty Advisor: Dr. Eric Cements

Fall 2005 Officers
President: Robyn Mainor
Vice-President: Travis Ratermann
Secretary: Diane Riley
Treasurer: Allison Marshaus
Historian: Catie Myers
Graduate Representative: Bill Gatlin
Faculty Advisor: Dr. Eric Clements

Spring 2006
President: Travis Ratermann
Secretary: Diane Sanders
Historian:
Faculty Advisor: Dr. Eric Cements

SEMO HP 25th Anniversary
Vice-President:
Treasurer:
Graduate Representative:

2006-2007
President: Diane Sanders
Secretary:
Historian:
Faculty Advisor: Dr. Eric Cements

Vice-President:
Treasurer:
Graduate Representative:

2007-2008
President: Sarah Stephens
Secretary: Megan Atchley
Historian: Michael Painton
Faculty Advisor: Dr. Joel Rhodes?

Vice-President: Elizabeth McGarvey
Treasurer: Mary Thrasher
Graduate Representative: Sam Sampson

2008-2009
President: Megan Atchley
Secretary:
Historian:
Faculty Advisor: Dr. Eric Cements

Vice-President:
Treasurer:
Graduate Representative:

2009-2010
President: Lea Bohr
Secretary: Lauren Pollien
Historian: Reid McCutcheon
Faculty Advisor: Dr. Eric Clements

Vice-President: LaDonna Garner
Treasurer: Stephanie Schwartz

2010-2011 Officers
President: Stephanie Schwartz
Secretary: Samantha Hennenfent
Historian: Andrew Scott
Faculty Advisor: Dr. Eric Clements

SEMO HP 30th Anniversary, celebrated 9 April 2011
Vice President: Chris Kinder
Treasurer: Rebecca Schmitt

2011-2012
President: Chris Kinder
Secretary: Jessica Reisenbichler
Historian: Brittany Contratto
Faculty Advisor: Dr. Eric Clements

Vice-President: Samantha Hennenfent
Treasurer: Rebecca Schmitt

2012-2013
President: Rebecca Schmitt
Secretary: Chelsea Edwards
Historian: Lindsey Lotz
Faculty Advisor: Dr. Eric Clements

Vice-President: Brittany Contratto
Treasurer: Samantha Kluesner

2013-2014
President: Chelsea Edwards
Vice-President: Lindsey Lotz
Secretary: Joe Abney
Treasurer: Dan Will
Historian: Ben White
Faculty Advisor: Dr. Eric Clements

2014-2015
President:
Vice-President:
Secretary:
Treasurer:
Historian:
Graduate Representative:
Faculty Advisor: Dr. Eric Cements

2015-2016
SEMO HP 35th Anniversary
President:
Vice-President:
Secretary:
Treasurer:
Historian:
Graduate Representative:
Faculty Advisor:Dr. Eric Cements

2016-2017
President: Rhiannon Martin
Vice-President:
Secretary:
Treasurer:
Historian:
Graduate Representative:
Faculty Advisor: Dr. Eric Cements

2017-2018
President: Rhiannon Martin
Vice-President:
Secretary:
Treasurer:
Historian:
Graduate Representative:
Faculty Advisor: Dr. Eric Cements

2018-2019
President: Kyle Petter
Vice-President:
Secretary:
Treasurer:
Historian:
Graduate Representative:
Faculty Advisor: Dr. Eric Cements

2019-2020
President: Tara Johnston
Vice-President: Alyssa Longman
Secretary: Jordan Cuneio
Treasurer: Madeline Van Leunen (FALL ONLY, left position)
Historian: Brent James (FALL ONLY, graduated)
Graduate Representative: -------
Faculty Advisor: Dr. Eric Clements

2020-2021
SEMO HP 40th Anniversary (delayed due to COVID-19)
President: Kaitlyn Barnett
Vice-President: Allyson Ingli
Secretary: Erika Kemp
Treasurer: --------
Historian: Connor Hundl
Graduate Representative: N/A
Faculty Advisor: Dr. Eric Clements

2021-2022

President: Kaitlyn Barnett
Secretary: Erika Kemp
Historian: Connor Hundl
Faculty Advisor: Dr. Eric Clements

40th Ruby Jubilee celebrated 13 November 2021

Vice-President: Allyson Ingli
Treasurer: ----------
Graduate Representative: N/A
Social Media: Faith Kinkelaar

Annual Banquets and Awards Ceremonies

The Historic Preservation Association (HPA) of Southeast Missouri State University in conjunction with the Historic Preservation Program hosts a banquet each year that includes a guest speaker and awards presentations (see below). This was the program for the HP Program's 25th Silver Anniversary Banquet:

We would like to invite you to celebrate the
25th Anniversary of the Historic Preservation Program
April 21, 2006 - April 23, 2006

Friday, April 21, 2006:
6:00pm-9:00pm
Reception at the Marquette Grand Lobby, located at 338 Broadway.
Appetizers and cash bar will be provided by Gatherings Cafe.

Saturday, April 22, 2006:
10:00am-11:30am
Walking Tour of Historic Downtown Cape Girardeau, provided by alumna Terri Foley
Please meet at the Cape Girardeau Convention and Visitors Bureau, 100 Broadway

1:00pm-3:00pm
Open House at the Marquette Center, formerly known as the Center for Regional History

1:00pm-3:00pm
Tours of the new Southeast Explorer (Traveling Museum), provided by alumna Robyn Mainor

1:00pm-3:00pm
Tours of the Academic Hall Dome, provided by members of the HPA

Banquet features Donovan Rypkema as keynote speaker[62]

Sunday, April 23, 2006:
11:00am-12:30pm
Flood Wall Mural Tour, provided by Dr. Frank Nickell

Awards

Arthur H. Mattingly Award in Historic Preservation

"The Historic Preservation Program, the Historic Preservation Association (HPA), the Center for Regional History, and the Department of History and Anthropology at SEMO are pleased to sponsor the Arthur H. Mattingly Award in Historic Preservation.... This award is given to a graduate of the Historic Preservation Program at Southeast, who has at least five years of experience in a related field, and who is demonstrating distinguished professional work that would be considered a credit to Southeast's program and to the profession."[63]

The award is named after the leading founder and proponent of SEMO's Historic Preservation Program.

Born in Southeast Missouri, Art Mattingly attended Southeast Missouri State College with a major in Social Studies Education. He went on to receive an MA in history from the University of Missouri and a Ph.D. from Kansas State University in 1971.

Though Mattingly listed thirty courses in the catalog he could teach, his favorites revolved around the American military experience. He was, in fact, selected to teach a workshop on military history at West Point.

His feeling that history should also be "hands-on" led to his leadership in the creation of the Historic Preservation Program at Southeast. This hands-on approach to historic preservation led to one of the most interesting projects undertaken in the program. In 1983 a historic log house was threatened with destruction. The owners were willing to donate the structure to the university if it were moved off site. Dr. Mattingly and his students presided over the dismantling and reconstruction of the Hunter House. Mattingly hoped that the reconstructed building could be the centerpiece for a living history site and this became a top priority for the department. Funding was never provided and the dream was never realized.

Mattingly was one of the first historians to realize that an important historical source was rapidly disappearing. Long before the present interest in World War II and the men who served, Art Mattingly started an extensive oral history project that would interview, tape, and catalog the experiences of area WWII veterans.

SEMO HP Program Faculty Bio, ca. 2018
semo.edu/faculty/emeritus.html (defunct as of 2021)

Dr. Mattingly who retired as a member of the in 1992 after 30 years of service, was director of the Historic Preservation program since its inception. Students created a memory scrapbook for "Doc's" entry into retirement . . . and emeritus.

Sandra Shupp-Campbell, when President of the Historic Preservation Association in 1994-1995, said, "The [Mattingly] Award is more than just doing your job; it is interacting with your community, sharing public history and being a leader." Dr. Mattingly added at the time, "It's an honor to be the award benefactor. It's nice to know that students that graduate and achieve some success in their chosen field of historic preservation receive an award for their accomplishments."[64]

When writing in 2014 for Cape Girardeau Central High School's blog "Cape Girardeau History and Photos: Coming of Age in Cape Girardeau," CGCH classmate of 1965 Ken Steinhoff wrote, "Arthur Mattingly brought history to life." He described his history professor to a friend while college: "He's talking when he walks into the room, and he's still talking when the bell rings and people are walking out. . . that was Arthur Mattingly, one of the best profs I had at SEMO or Ohio University."[65]

A plaque with past awardees is on display in the Department of History and Anthropology's offices.

Drs. Frank Nickell and Art Mattingly, 25th Silver Anniversary Annual SEMO HPA Banquet, 2006

Arthur H. Mattingly Award in Historic Preservation Recipients

Mattingly Awardee	Graduated	Award Year
Jesse "Jess" Francis	1983	1992-1993
Joan (Tinsley) Feezor	1982	1993-1994
Richard Lloyd "Rick" Young	1982	1994-1995
Debra Ann Reid, Ph.D.	1982	1995-1996
Susan Marie "Susie" Haake	1989	1996-1997
Brian Keith McCutchen	1992	1997-1998
Jacqueline A. "Jackie" (Holt) Jefferson	1986	1998-1999
Renee Elizabeth Bomgardner	1994	1999-2000
Stephen C. Rogers	1994	2000-2001
Michelle Christine (Saxman) Rogers	1995	2000-2001
Jon Gregory Colburn	1995	2001-2002
Stacy Ann (Webb) Reaves	1995	2002-2003
David Wayne Jackson	1993	2003-2004
Karen L. Daniels	1988	2004-2005
Heather Brianna (Engelmann Farrow) Carey	2001	2005-2006
Dan Schoeneberg	1995	2006-2007
Bridgette Lee (Epple) Kelch	2000	2007-2008
John Eric Bry	1995	2008-2009
Danielle Lewon	1996	2009-2010
Carla L. (Taylor) Jordan	2004	2010-2011
Lesley L. (Lehr) McDaniel	2000	2011-2012
Ellen M. Ryan	2006	2012-2013
Donna Jean Rausch	2000	2013-2014
William Michael "Bill" Hart	1997	2014-2015
Rebecca L. "Becky" (Rolwes) Sawyer	1996	2015-2016
Melissa A. (Dirr) Gengler	1990	2016-2017
Travis Allen Ratermann	2007	2017-2018
Garret B. Kremer-Wright	2005	2018-2019
NONE, DUE TO THE COVID-19 PANDEMIC		2019-2020
Denise Morrison	1982	2020-2021

HPA Spirit Award

The HPA has presented *most years* since 1994 an *HPA Spirit Award.* This recognition is bestowed upon one or two outstanding HPA members who have exemplified and personified the spirit of the HP Program's mission and goals during their SEMO student career.

Three plaques with past honorees of the *HPA Spirit Award* are on display in the Department of History and Anthropology's offices, recognizing:

1994 Felix Kinsley
1995 Ken Smith
1996 Becky Bradley
1997 Joe Barber
1998 Will Johnston
1998 Chadwick Boehlke
1999 Leslie Dingman
1999 Rich Jackson
2000 Bridgette Epple
2001 Jeremy Wells
2002 Jennifer Brewer
2003 Robyn Mainor
2005 Eric Clements
2006 Diane Riley
2007 Edward Farrell
2011 Samantha Hennenfent
2012 Rebecca Schmitt
2013 Chelsea Edwards
2014 Riley Price
2015 Daniel Will
2016 Rhiannon Martin
2017 Chelsi Montgomery
2018 Katy Armstrong
2019 Jordan Cuneio
2021 Faith Kinkelaar

Faculty Merit Award

Dr. Frank Nickell, was awarded with the University's most prestigious award, the Alumni Association Faculty Merit Award, in 1991-1992. His photograph joins others in Kent Library.

Up to that time, Dr. Charles "Charlie" Sharp was the only other member of the History and Anthropology Department to have been so honored (in 1984), according to the 17th edition of *Pastscript,* 1991-1992.

In 1999, Mr. Larry Easley received the award.

And, in 2021, Steven Hoffman received the Faculty Merit Award.

Young Alumni Merit Award

This award, presented by the Alumni Association from the College of Liberal Arts, has gone to three HP alumni as of 2021:

Susan Haake, 1989 | John Bry, 2002 | Debra Reid, Ph.D. 2020

Honor Society Chartered

SEMO's Epsilon Chapter of Sigma Pi Kappa the international society for historic preservation students and faculty, was chartered on March 23, 1996. Sigma Pi Kappa is the only international society specifically for historic preservationists. The Greek letters Sigma Pi Kappa represent Scholar, Protector of Cultural Resources.

By 1995-1996, five graduate and ten undergraduate students had been inducted. Dr. Mattingly was awarded an honorary membership at the Spring 1996 HPA banquet. Graduates of the HP program with an overall undergraduate GPA of 3.5 (3.7 for graduate students) are invited to apply.

The Chapter has presented an "Excellence in Historic Preservation Award." A future edition of this retrospective may include more about past years' recipients. One Cape Girardeau resident, Barbara Kaye Smith Hamblin, received the award "for the restoration of the historic home at 313 Themis Street in Cape."[66]

In 2021, the Epsilon Chapter recognized its 26th class for the 40th year of the SEMO HP Program

8

HP Program Coordinators, HP Faculty and Interdisciplinary Professors

Across its first 40 years, the SEMO HP Program has been guided by four industrious and enthusiastic coordinators as well as an unnumbered corps of supporting faculty both within the Department of History and Anthropology and across SEMO's academic colleges.

HP Program Coordinators

Dr. Arthur H. Mattingly, founder and coordinator of the Historic Preservation Program and later faculty advisor to the Historic Preservation Association, announced in Spring 1992 that he would, after more than a decade of involvement in the HP Program, be taking early retirement at the end of the coming fall semester.[67] He continued to teach one course per semester through 1993-1994.[68] Mattingly was awarded Emeritus status in the Department of History and Anthropology for his 40 years of distinguished service between 1963 and 1992.[69]

The Department of History and Anthropology hired Dr. Richard Betterly, D.A., from Middle Tennessee State University, to replace Mattingly as Coordinator of the HP Program and Assistant Professor of History.[70]

Soon thereafter, Betterly left SEMO and Dr. Bonnie Stepenoff was named coordinator of the Historic Preservation Program in Spring 1995. She held her doctorate in

American History from the University of Missouri-Columbia, and had background in cultural resource preservation, acquisitions, and archival administration.

Dr. Steven J. Hoffman, lecturer at Bradford College in Bradford, Massachusetts, joined the department as an Assistant Professor in Autumn 1995. He earned his Master of Heritage Preservation degree from Georgia State University and his Ph.D. from Carnegie Mellon University. Hoffman was Associate Professor from 2001 and 2006, when he was promoted to Professor. Meanwhile, in 2003, Hoffman also accepted the position of Coordinator of the Historic Preservation Program, where he continues through 2021. In August 2021, he became the Director of the Center for Regional History. The author wishes to thank Dr. Hoffman for editing and contributing key points to this retrospective.

Historic Preservation Program Faculty

While recognizably incomplete (esp. in a first edition of a compilation of this nature), acknowledgement must be extended to some of the many professors who have led HP students towards their unique degree in historic preservation.

First, a core teaching faculty within the Department of History and Anthropology have been officially "HP Program" instructors, including but not limited to:

Dr. Richard Betterly
Dr. Eric Clements
Dr. Steven Hoffman
Dr. Arthur H. Mattingly (Professor Emeritus, 1963-1992)
Dr. Joel Rhodes
Dr. Lily Santoro
Dr. Bonnie Stepenoff (Professor Emeritus, 1993-2012)
Dr. Jane Stephens (Professor Emeritus, 1978-1994)
Mr. Bob White (Professor Emeritus, 1966-1999)

Department of History and Anthropology Faculty Offering Historic Preservation Classes[71]

Secondly, Department of History and Anthropology faculty have taught classes—including some curriculum with core "HP" designation—but never likely considered themselves to be "HP Program Faculty," nor are/were they technically or officially recognized by SEMO as HP Program professors. Yet, their service and contributions to the HP Program are/were vital to its success. These professors may include but are not limited to (individuals in **bold** type were instrumental since the beginning of SEMO's HP Program in 1981):

Dr. Toni Alexander *(Chair, Department of History and Anthropology)*
Dr. Warren Anderson, *Anthropology/Archeology*
Dr. Jennifer Bengtson, *Anthropology/Archeology*

Dr. Adam Criblez
Dr. Daryl Fridley
Dr. Vicky McAlister
Dr. Carol Morrow, *Anthropology/Archeology*
Ms. Jennifer Rickard

Professor Emeriti (by retirement year)
Dr. E. Eugene Nutter (1955-1980)
Dr. Harold H. Dugger (1960-1981)
Dr. George A. Ketcham (1959-1990)
Dr. Charles C. Bonwell (1968-1993)
Dr. George G. Suggs, Jr. (1964-1995)
Dr. Ernest K. Farmer (1967-1996)
Dr. Gene Ramsey (1967-1997)
Dr. Truman H. Smith (1965-1997)
Dr. John P. Coleman (1967-2001)
Dr. Martin D. Needels (1969-2001)
Dr. Charles E. Sharp (1966-2001)
Dr. Robert H. Skelton (1970-2001)
Dr. Alberta L. Dougan (1972-2006)
Mr. Larry J. Easley (1967-2007)
Dr. James Christopher Schnell (1970-2009)
Dr. Joseph R. Werne (1972-2011)
Dr. Frank Nickell (1969-2013)[72]
Dr. Dalton Curtis (1979-2017)
Dr. Warren Anderson (1997-2018)

Interdisciplinary Faculty

Finally, yet no less importantly, there have been unnumbered faculty across SEMO's many "colleges" of study who have taught interdisciplinary courses to HP students.

As stated before, the Historic Preservation Program was SEMO's first degree program to incorporate interdisciplinary curriculum.

This debut list is only our first effort as we begin to assemble an honorary enumeration of professors in interdisciplinary areas that include:

Ms. Beth Adkins, *Geography/Cultural Geography*
Mr. Bryan Bowers, *Construction Management/Architectural Design*
Dr. Indi Braden, *GIS*
Dr. Chamberlain, *Art*
Dr. Joni Hand, *Art History*
Dr. Hogan, *English and Professional Writing*
Dr. Maryann Kellerman, *Interior Design*[73]
Ms. Carol Nesler, *Interior Design*
Dr. Michael Roark, *Cultural Landscapes*
Dr. Art Rosser, *Applied Arts/Industrial Technical Education*
Ms. Morgan Schmid, *Interior Design*

Taking Interdisciplinary to Infinity . . . and Beyond

Drs. Maryann and Bert Kellerman, longtime SEMO Professors, have and continue to put their time, talent, and significant tithe towards historic preservation at SEMO, in Cape Girardeau, and throughout southeast Missouri. They own and have restored SEVERAL significant properties, and established the Kellerman Foundation for Historic Preservation, "to encourage and support HP in southeast Missouri and southern Illinois, while connecting the community to its architectural, historical and cultural legacy by telling its stories." Hats off to the Kellermans who have also created the Nickell Regional Heritage Archive to document, archive, digitize, and distribute the 50-plus year collection of longtime SEMO Professor, Dr. Frank Nickell.

9

SEMO HP Alumni Chapter

A handful of former graduates of the SEMO HP Program had for years expressed an interest in "doing more" for their alma mater. On November 19, 2014, a Facebook page was created. Then, in 2016, Sandra Shupp-Campbell (BSHP 1996) helped organized the alumni weekend for the program's 35th anniversary. She attracted at that time the dedication of a handful of other graduates who formulated the SEMO HP Alumni Chapter.

After months of planning and coordinating with the SEMO Alumni Foundation, the Alumni Association Board of Directors sanctioned its *first-ever degree-specific alumni chapter* bylaws in October 2016. In fact, five years later in 2021, the SEMO HP Alumni Chapter remains the only such chapter of its kind at SEMO.

The alumni chapter's website (**semohpalumni.org**) posts current and long-term, historical data, with a members-only section. And the chapter's Facebook page (**facebook.com/semohpalumni**) has 247 followers/236 likes as of October 2021.

Membership is optional. Dues are $10/year to help defray administrative costs such as hosting a url and the website, postage, and incidentals. Some alumni pay for multiple years of membership at once. Whenever the chapter's fund allows, excess revenues are transferred to a particular fund or project for SEMO HP students' benefit, as coordinated through the HP Program Coordinator.

The volunteer officers of the chapter dedicate themselves to the four main purposes stated in the Constitution and Bylaws:

ARTICLE I: NAME & AFFILIATION

Name – The name of this chapter shall be the "Southeast Missouri State University Historic Preservation Alumni," an affinity alumni chapter under the auspices of the Southeast Missouri State University Alumni Association.

Institutional Affiliation – The rules and regulations of the Alumni Association shall be followed when not inconsistent with the rules and regulations of Southeast Missouri State University Historic Preservation Alumni.

ARTICLE II: MISSION & PURPOSE

Mission – The mission of this chapter shall be to build valued and lasting relationships with former students, graduates and friends of Southeast Missouri State University's Historic Preservation Program.

Purpose – The chapter seeks to accomplish its mission by the following means:

1. Share their experiences and donate archival materials to Southeast Missouri State University Special Collections and Archives documenting their tenure with the Southeast Missouri State University Historic Preservation Program;

2. Usher current Southeast Missouri State University Historic Preservation students and new Southeast Missouri State University Historic Preservation graduates into the 'real world;'

3. Connect with alumni members personally and professionally; and,

4. Encourage each alumni to support and contribute to the Historic Preservation Fund for Experiential Learning, so that once endowed, will directly benefit the Southeast Missouri State University Historic Preservation Program by assisting students with hands-on experiences outside of the traditional classroom.

ARTICLE III: MEMBERSHIP ELIGIBILITY, DUES & DONATIONS

Membership Eligibility - Membership shall be open to current Southeast Missouri State University Historic Preservation program students; former Southeast Missouri State University Historic Preservation program students; Southeast Missouri State University Historic Preservation graduates; Southeast Missouri State University Historic Preservation program founders, coordinators, and educators; and, friends who otherwise support the chapter's mission.

Dues – Dues shall be solicited, encouraged, and directed to a ***non*-tax-deductible** account, managed by the Southeast Foundation, which will be dedicated/restricted to support nominal operating expenses of Southeast Missouri State University Historic Preservation Alumni (i.e., website; bank account; letterhead, envelopes, and, postage; and, other nominal operating expenses to aid the chapter in fulfilling its mandatory operational obligations). Dues will NOT be tax deductible. The minimum, annual dues will be $10, and may provide members access to a "members only" portion of the chapter's website. Funds collected will help defray the operating costs of the chapter, as above stated; and,

Donations – Donations shall be solicited, encouraged and directed to a **tax-deductible** "Historic Preservation Fund for Experiential Learning," managed by the Southeast Foundation, with the specific goal to endow the fund at least to $10,000 so that the Southeast Missouri State University Historic Preservation Program may begin utilizing realized accrued interest. Donations to the Historic Preservation Fund for Experiential Learning ARE a tax-deductible contribution. The chapter will instruct giving directly to the Historic Preservation Fund for Experiential Learning when they wish to give a tax-deductible contribution. A web link will promote the Historic Preservation Fund for Experiential Learning, and also provide a form that can be printed-off and sent to the Foundation via the USPS.

ARTICLE IV: OFFICERS

Officers – The officers of the chapter shall be a President, Vice-President, and Secretary, elected by the membership.

Liaison – The current year's President of the Historic Preservation Association (HPA) student group may act as liaison between current student body and Southeast Missouri State University Historic Preservation Alumni chapter membership, to encourage student body to participate in annual event(s), and to join the chapter's membership.

Eligibility – Officers must be former Southeast Missouri State University Historic Preservation Program students and/or graduates.

Election – The officers shall be elected by members in attendance at the chapter's annual event through a hands-up/hands-down popular vote. Nominations may be accepted from the floor.

Term – The officers shall serve from the date of election until the date of the next election; generally, a one-year-term. There are no term limits, as long as officers are willing to serve, and are duly elected.

Vacancy – If a vacancy occurs, the remaining officer(s) may appoint a successor to serve until the next election. Ideally, the chapter should strive to have at least one officer who has tenure for organizational memory and smooth succession.

ARTICLE V: DUTIES OF OFFICERS

President – it shall be the duty of the President to:

- Liaison with the Alumni Association and HP Program Coordinator;
- Preside at meetings;
- Vote only in case of a tie;
- Represent and promote the chapter;
- Direct the chapter on tasks or projects;
- Post to the chapter's Facebook page, and any other social media employed;
- Arrange at least one annual event;
- Appoint a Vice President, in case of a mid-term vacancy;
- Appoint volunteers to serve and accomplish a specific task or project; and,
- Report revenues and expenses through the Foundation; and,
- Perform such other duties as ordinarily pertain to this office.

Vice-President – It shall be the duty of the Vice-President to:

- Preside at meetings, in the absence of the President;
- Represent and promote the chapter;
- Post to the chapter's Facebook page, and any other social media employed;
- Issue notices to membership of at least one annual event;
- Assume the Presidency and/or appoint a President, in case of a mid-term vacancy;
- Work with the President to appoint volunteers to serve and accomplish a specific task or project.

Secretary – It shall be the duty of the Secretary to:

- Perform any secretarial duties;
- See that any non-current records are transferred to the chapter's archival collection in the Southeast Missouri State Special Collections and Archives;
- Encourage and assist the HPA and Historic Preservation Program to archive its non-current historical records to the appropriate collection(s) in the Southeast Missouri Special Collections and Archives;
- Maintain a roster of the chapter, and post-it, and any other pertinent files to the chapter's website; and,
- Maintain the chapter's website (unless delegated to a volunteer); and, post to the chapter's Facebook page, and any other social media employed.

ARTICLE VI: MEETINGS

Meetings – Regular Executive Board meetings between all Officers shall be held semi-annually, and may be teleconferenced or via other electronic/telephonic/virtual format. At least one annual membership meeting for the chapter's constituents shall be held, preferably in Cape Girardeau, and preferably coinciding with the HPA's annual event each spring.

Special Meeting – Special meetings may be called by the President at any time, and may be teleconferenced or other electronic/telephonic/virtual format.

Quorum – Understanding that the membership is geographically disbursed and generally unable to attend regular meetings, a quorum at any annual event where elections are held shall consist of 1/10 of members in good standing. Regular meetings must include all Officers, or their proxy vote via electronic/telephonic/virtual presence.

Parliamentary Authority – Robert's Rules of Orders, shall govern this chapter in all cases to which they are applicable and in which they are not inconsistent with these bylaws.

ARTICLE VII: COMMITTEES

Committees – Committees may be formed at the President's request and appointment, and include volunteers interested and competent to accomplish a specific task or project.

ARTICLE VIII: AMENDMENTS

Selection – These bylaws may be amended by a majority vote of the chapter membership in attendance at an annual event where elections and/or other business is conducted.

Notice – All members shall receive advance notice of the proposed amendment(s) in any format deemed appropriate by the officers, at least five days before the annual meeting.

SEMO HP Alumni Chapter Officers

2016-2017
President: Sandra (Shupp) Campbell (BSHP 1996)
Vice President: Melissa (Dirr) Gengler (BSHP 1990; MAHP 1992)
Secretary: David W. Jackson (BSHP 1993)

2017-2018 through 2020-2021
President: Melissa (Dirr) Gengler (BSHP 1990; MAHP 1992)
Vice President: Jon Colburn (BSHP 1994)
Secretary/Webmaster: David W. Jackson (BSHP 1993)
Social Media Chair: Sandra (Shupp) Campbell (BSHP 1996)

Melissa (Dirr) Gengler received her Bachelor's degree in Historic Preservation in 1990 and her Master's degree in history in 1992. She worked for both the South Dakota and Nebraska State Historic Preservation Offices and in 2010 left the public sector to open her own consulting company called Historic Resources Group, Inc. (hrg-nebraska.com). Since 2010, Melissa has been working on recordation and compliance projects for both public and private sector clients. She returned to SEMO for the 35th Anniversary celebration and has attended every year since. She is a founding member and serves as President of the SEMO HP Alumni Chapter (**semohpalumni.org**).

Sandra (Shupp) Campbell received her Bachelor's degree in Historic Preservation in 1996. She resides in Murfreesboro, TN, with her husband, David. Sandra is interested in local history and engages in research projects with the county historical society, preservation groups, and her colleagues where she works at Middle Tennessee State University. Sandra spearheaded the development of the SEMO HP Alumni Chapter and hopes all graduates will support the HP Program and the Chapter.

David W. Jackson graduated magna cum laude with a BS in HP from SEMO in 1993. He enjoyed a 20+year career as archivist for Unity School of Christianity and Jackson County (Mo.) Historical Society, both in the Kansas City, MO, metro area. Since 2000, David has—on his own or in concert with others—added more than 40 titles to local history bookshelves. He continues to research, write, publish, and present on local history and its preservation through his publishing house, The Orderly Pack Rat (orderlypackrat.com), which he founded in 1996.

Jon Colburn, SEM O BSHP 1995, Horticulturist and Historic Preservationist is a Facilities Specialist with La Roche University, in Pittsburgh, PA. Active in preservation he has worked at The Hackley & Hume Historic Site in Muskegon, MI and at The Frick - Pittsburgh, served as an Historic District Commissioner, and served on boards of several preservation associations. When he takes time from duties with Girl Scouts Western Pennsylvania and at Smithfield United Church of Christ, Colburn, his husband David, and their chiweenie, Izzy, enjoy watching scullers in the Ohio River from their Neville Island home.

10

Testament to a Legacy

The legacy of the SEMO HP Program is grounded in the financial resources that benefactors have invested to ensure the program endures and thrives. The interest that a principled account offers students in need is quantified by the trajectory their careers take because of the enhanced education they receive at SEMO. And prospective HP students are drawn and attracted from across America. For instance, Terri Foley enrolled at Southeast because of its preservation program. "It was the best rounded program, offering classes in archaeology, historic sites, museums, archives and architecture," she said. Foley graduated from the SEMO HP Program in 2002 and when interviewed in 2006, operated an historic preservation consulting business in Cape Girardeau.

> All individuals who have graciously bequeathed funding for the funds, scholarships, and other endowments below are recognized for their generosity.
> Your selfless dedication is admired and is to be emulated.
>
> Please accept an eternal "thank you" on behalf of past, present, and future SEMO HP students who have, are, and will be inspired and benefit from your selfless gifts.

But the benefits of the SEMO HP Program go much further.

The SEMO HP Program's success is also visible through the improved beautification and creation of historic districts around Cape Girardeau, Missouri, as well as the SEMO campus, and radiate to countless communities across America. That is a legacy.

Way back in 2006, Dr. Mattingly and Dr. Nickell were interviewed by the *Southeast Missourian* on the occasion of the 25th anniversary of the HP Program they co-founded. Writer Mark Bliss recalled of 1981, "Back then, most people saw worn-out buildings as eyesores to be torn down rather than architectural gems in need of care and attention. But

Mattingly and Nickell . . .said the view of the public as well as city and civic leaders has changed dramatically. What was once a novel idea has prospered." That is a legacy.

The SEMO HP Program has graduated and released into communities across the nation more than 350 dedicated individuals who work at museums, historic sites and other preservation-related careers . . . and beyond! That is a legacy.

Bliss continued, "The program also has helped create a culture of preservation in the Cape Girardeau area and has helped spark building restoration projects in the city's downtown, say those connected with the program." Dr. Steven Hoffman, HP Program Coordinator since 2003 added, "Of the 23 individual structures in Cape Girardeau on the National Register of Historic Places, 14 have been listed by historic preservation students, faculty and alumni. Besides the availability of preservation students and graduates, building restorations locally have been sparked by state and federal tax credits that have made preservation financially feasible," Hoffman said. And "Historic preservation is getting much more attention than ever in Cape Girardeau," said Foley. "There is an awareness of what the National Register is in town because of the program," she added. That is a legacy.

It gets better. "Historic preservation has helped generate tourism for Cape Girardeau and other Southeast Missouri communities," Hoffman said, "because most people learn the history of an area by visiting historic homes, sites and museums."

Mattingly summarized the legacy he takes pride in. "These kids are going to go out there like missionaries and work for preservation."

And the best legacy that these 'missionaries' . . . ambassadors . . . graduates . . . alumni . . . can do is to give back to the SEMO HP program *somehow*—whatever creative way they are capable—to safeguard the continued success of the program that attracted them to SEMO in Cape Girardeau.

Make the Mount Vernon Ladies proud!

Mount Vernon Ladies Association
clements.umich.edu

Show SEMO the Money

Now to the brass tax. The nitty gritty. It's time to "Show SEMO the money."

SEMO's Historic Preservation students have access to a variety of resources for financial aid, the most notable of which are scholarships processed through the Bursar's Office. These are available because of the generosity of many who have come before and made significant financial pledges to provide aid to HP students.

Department of History and Anthropology Scholarship Funds Available to HP Majors

The list that follows are scholarships that HP students may apply for to help them with their tuition or educational expenses. *Whenever possible, historical factoids on each fund and/or its generous benefactors are included in italics.*

Lawrence and Alice Breeze Scholarship in History

Southeast Missouri State University Graduate Program students pursuing a Master of Arts in Public History or A Master of Arts in History. Financial need is a consideration, but a FAFSA form is not required.

Dalton "Rusty" B. Curtis, Jr. Scholarship

The scholarship will be awarded to an undergraduate student attending Southeast Missouri State University majoring in History, Historic Preservation, or Social Studies Education with a 3.25 or greater GPA> Financial Need is preferred not required.

Alberta Macke Dougan Graduate Scholarship in History

The recipient of this scholarship will be a graduate student enrolled at Southeast Missouri State University in History. This scholarship will be given to a graduate student or graduate students completing an internship or advanced project in applied history, or engaging in or presenting original research or other scholarly or experiential activity.

Dr. Harold Dugger Endowed Scholarship

The scholarship will be awarded to a full-time undergraduate student attending Southeast Missouri State University majoring in a program administered by the Department of History, with a minimum cumulative GPA of 3.0. Preference will be given to first-generation or nontraditional students. Financial need will also be considered but not required.

Elgie Woods Foster Daughters of the American Revolution Scholarship

Open to undergraduate students with 15 credit hours and a cumulative 2.5 GPA. Applicant must be enrolled in History or the Historic Preservation program with an emphasis in local/regional history and have a FAFSA on file. A separate application can be obtained from the History Department Office in Carnahan, Room 311. Applications must be returned to the History Department by March 1st.

Historic Preservation Experiential Learning Scholarship

A $10K endowed fund established in 2019 by Dr. Debra Reid, one of SEMO HP Program's first graduates in 1982.

The recipient of this scholarship will be undergraduate students majoring in Historic Preservation or graduate students majoring in Public History with an emphasis in Historic Preservation. Recipients need to have completed or be currently enrolled in experiential learning opportunities including, but not limited to, internships, applied projects, or field school.

History Graduate Student Endowed Scholarship

The recipient of this scholarship will be a graduate student enrolled at Southeast Missouri State University in History. This scholarship will be given to a graduate student or graduate students completing an internship or advanced project in applied history, or engaging in or presenting original research or other scholarly or experiential activity. The scholarship will be awarded by the departmental scholarship committee.

William T. Jenkins Memorial Scholarship

Established in the memory of Dr. William T. Jenkins, Professor of History at SEMO from 1966-1974.

Open to University juniors pursuing a B.A. in History, a B.S.E. in Social Studies, or a B.S. in Historic Preservation, with an overall 3.0 GPA and no grade lower than a "B". Applicant must submit a current transcript and three letters of recommendation focusing on his or her potential for academic success. Selection is made solely on the basis of merit. A separate application can be obtained from the History Department Office in Carnahan, Room 311. Applications must be returned to the History Department by March 1st.

Lawrence and Donna Marler Endowed Scholarship

Open to Historic Preservation majors with a 2.5 GPA. Preference will be given in the following order to high school graduates of the City of Ste. Genevieve, Ste. Genevieve County, Wayne County, Perry County, St. Francois County, the Southeast Missouri region, and the State of Missouri. Deadline of March 1st. A separate application can be obtained from the History Department Office in Carnahan, Room 311. Applications must be returned to the History Department by March 1st.

Mulkey-White Preservation Scholarship

To assist a second-year major on the basis of academic performance and financial need. Patty Mulkey, a graduate of SEMO, taught in Southeast Missouri schools for 30 years, and was involved in numerous historic preservation activities including serving as

President of the Cape River Heritage Museum and Direct of Tours and Volunteers of the Historical Association of Greater Cape Girardeau. Billy "Bob" White, a member of SEMO's History Department for 25 years (as of 1994-1995), served as Director of the Center for Regional History and the Cape Historic Preservation Commission.

Open to second-year Historic Preservation majors who maintain a minimum 3.0 GPA. A separate application can be obtained from the History Department Office in Carnahan, Room 311. Applications must be returned to the History Department by March 1st.

Milton H. & June E. Nothdurft Memorial Restricted Scholarship

The recipient of this scholarship will be a student enrolled at Southeast Missouri State University majoring in Historic Preservation. First preference will be for a graduate of Jackson High School in Jackson, Missouri.

Old Town Cape Dr. Steven J. Hoffman Historic Preservation Scholarship

The recipient of this scholarship will be a student enrolled at Southeast Missouri State University. First preference is for a student with a major or minor in historic preservation with a demonstrated interest and passion for historic preservation based on economic revitalization. Second preference is for a student with a major or minor in historic preservation and third preference for a student with a major in the Department of History.

River Bend Friends of Historic Preservation

Open to Historic Preservation majors with a "B" average who are dedicated to their field. First preference will be given to juniors, and second preference will be given to seniors. Applicants must obtain and return applications to the History Department.

W.H. and Sarah Bowers Zinn Memorial

Applicants must be from the Southeast Missouri area, be majoring in Historic Preservation, and have a minimum GPA of 3.0. A separate application can be obtained from the History Department Office in Carnahan, Room 311. Applications must be returned to the History Department by March 1st.

Faculty are the beneficiary of endowed chairs.

Investing in History and HP FACULTY

Endowed Chairs

*Awarded to deans, department chairs, and superior faculty to enhance their scholarly efforts and research. Minimum funding required to establish an endowed chair is **$1,000,000.***

Endowed Professorships

*Awarded to outstanding senior faculty members to reward and enhance outstanding teaching and scholarly accomplishments. Minimum funding=**$500,000.***

Distinguished Faculty Fellowships

*Support senior faculty recognized for excellent teaching and outstanding scholarship. Minimum funding=**$250,000.***

Faculty Fellowships

*Support faculty demonstrating very effective teaching and scholarly accomplishments, regardless of rank or tenure. Minimum funding=**$100,000.***

Faculty Excellence Funds

*Dedicated to promoting faculty teaching, scholarly work and continuing education. Funds can be used to respond quickly to opportunities that arise unexpectedly and the potential for having a positive impact on the College's programs and students. Minimum funding=**$20,000.***

Other sources fund HP Program activities for enrolled students' benefit. Endowed funds and membership to the SEMO HP Alumni Chapter are serviced through the Alumni Foundation, and are an excellent way for alumni to give back to their alma mater

.

Endowed Funds

Each fund below has been endowed to at least $10,000 minimum.

The HP Program Coordinator may draw from the interest of these accounts to assist with current students' experiential learning.

Anyone may add to these accounts by making a tax-deductible donation at any time. As the principal grows, more interest may become available for students' benefit. Regrettably, at the time of this first-edition printing of the SEMO HP Program's retrospective, more detailed information about the origin of many of these funds and their benefactors was not readily available.

HP Fund

A general endowed fund created in the 1990s to enhance students' experiential learning.

Lawrence and Alice Breeze Library Endowment Fund

Dr. Arthur H. Mattingly Endowed Historic Preservation Fund for Student Experiential Learning

"Dr. Arthur and Nancy Mattingly of Cape Girardeau made a $10,000 gift to the Foundation to establish the endowment to provide financial assistance to students in the Historic Preservation program for experiential learning. These opportunities include unpaid or underpaid internships, conducting research and implementing projects that relate to the built environment, museum studies, historic site administration and archives, as well as participating in professional conferences and workshops to enhance their educational experience and more.

"Dr. Mattingly was born in St. Louis, but spent his childhood in Charleston, Missouri. He joined the U.S. Air Force in 1954, serving three and a half years. He was discharged with the rank of staff sergeant and entered SEMO in 1958.

"Dr. Mattingly's personal goal of wanting to serve and help others, and his deep interest in history led him to a Bachelor of Science in Education in 1961. Not satisfied, and still eligible for the G.I. Bill, he went to the University of Missouri where he earned a Master of Arts in history in 1962.

"In the fall of 1962, he started teaching high school social studies at the North Kansas City High School, North Kansas City, Missouri. While awaiting the start of school in 1963, he was offered, and accepted, a position in the Department of History at SEMO. In 1964, in addition to his teaching, Dr. Mattingly was appointed the coordinator of the American History program serving until 1968, when he applied for a leave of absence to pursue a doctorate in history.

"He was accepted into the Kansas State University graduate program, won a graduate teaching assistantship in the History Department and started classwork and teaching in the fall of 1968. He completed coursework, exams and his dissertation, receiving his doctoral degree in the summer of 1971.
Dr. Mattingly then returned to Southeast teaching new history courses and publishing several articles. He completed the hundred-year history of Southeast Missouri State University, *Normal to University: A Century of Service,* in 1979.

"With his interest in local history, Dr. Mattingly was appointed director of the newly created Regional History Center in 1978, the Department of History's outreach program that included academic work on local and state history to broaden student views of history nearby. From 1979-1981, he advocated for and wrote with others a new degree, the Bachelor of Science in historic preservation. Upon approved by the Missouri Board of Higher Education in 1981, this program at Southeast was only the fourth such undergraduate program in the United States. The new degree was unique in its structure and focus as it was designed to teach history outside the classroom. Dr. Mattingly was chosen the program coordinator, serving in this capacity until retiring in 1993, after 30 years of learning from his students and sharing with them his humor and love of history."[74]

Harold Dugger Lecture Fund

Undergraduate Student Development Fund

Graduate Student Development Fund

Endowed after an initiative of University President Kenneth Dobbins.[75]

Faculty Development Fund

11

In Memoriam

Karen Joy Grace

Karen was born January 13, 1944, in St. Louis, a daughter of John W. and Edna Mae Nobles Ridlen. She was married January 10, 1960, in St. Louis, to Martin Wayne Grace Sr., who survived in Olive Branch, Illinois. She graduated from Southeast Missouri State University, earning a bachelor's degree in education, and a master's degree in history. She was a member of Phi Alpha Theta Honors Historical Society. Karen was an American history instructor at SEMO from 1981-1985.

For more than 15 years between 1985 and 2003, Karen was a cultural resource preservationist with the Missouri Department of Natural Resources. During that time, she coordinated education and outreach programs for the State Historic Preservation Office. A tireless advocate for preservation, she served on the Missouri Main Street Advisory Board, Lewis and Clark Commission, and the board of the Midwest Organization for the Recognition and Recording of Ethnic Heritage (MORARE). During her career, she provided tax credit review and technical assistance to property owners, supported Women's History Month programs, worked with Historic Schools Project, and developed Missouri's heritage curriculum. She also worked on: the Archaeology Challenge; Underground Railroad; Harrisonville Courthouse Square Project; Women's History Month; Missouri State Fair projects; tax credits review technical assistance; Historic Schools Project; Missouri Heritage curriculum videos; and preservation issues. She was an honorary member of the Jefferson City Garden Club.

Karen died at age 59 on March 11, 2003, at Villa Marie Skilled Nursing Facility. She was survived by her mother. Other survivors included: one son, Martin W. Grace Jr., Cape Girardeau; one daughter, Stephanie J. McCutchen, Cape Girardeau; one brother, Roger W. Ridlen, Fair dealing; three grandchildren; two step grandchildren; and two step great-grandchildren. She was buried in Cape Girardeau Memorial Park Mausoleum.[76]

Lucas Barton Greenwalt

Lucas Barton Greenwalt, 28, of Milwaukee, Wisconsin, formerly of Woodlawn, Illinois, passed away at 3:40 pm July 7, 2020, at his residence. He was born December 10, 1991, in Mount Vernon, Illinois to Bart and Angie (Dickey) Greenwalt.

In addition to his parents, Lucas is survived by his grandparents, Bennie and Donna Greenwalt of Wayne City, Illinois; and several aunts, uncles, and cousins. Lucas was preceded in death by his grandparents, Leland and Wilma Dickey; and uncle, Jerry Dickey.

Lucas was a member of Woodlawn Christian Church. He was passionate about playing the violin and loved his family very much. Lucas will be greatly missed by those who knew and adored him.

A Visitation will be held Monday, July 13, 2020, from 12:00 pm to 2:00 pm at Archie Sanders Memorial Chapel at Thomason Cemetery in Wayne City, Illinois. A Funeral Service will begin at 2:00 pm with Pastor Jerry Wilson officiating. Interment will follow at Thomason Cemetery in Wayne City. Memorials may be made in Lucas' honor to WisHope in Waukesha, Wisconsin.[77]

Lucas graduated with a BSHP from SEMO in 2015.

Thomas Ellis Holloway

Tom, of Lampe, Missouri, and formerly of St. Joseph, Missouri, died at age 69 on May 30, 2012, at Cox South Hospital, Springfield, Missouri. He was born January 15, 1943, to Chester A. and Estelle (Plante) Holloway in Barron, Wis. He married Janey Glick in 1987.

Tom graduated from Barron High School and attended the University of Wisconsin, River Falls. He graduated from Brown Institute in Broadcast Journalism and later from Southeast Missouri State University in Historic Preservation. He thoroughly enjoyed driving a tour bus and entertaining his passengers with information about surrounding sites, enabling him to use his broadcasting and history background. He loved history and sharing his knowledge with others.

Tom was a Christian man who attended several churches in his lifetime, attending most recently, the Tabernacle in Branson, Missouri.

He was preceded in death by his parents; and brother, David M. Holloway. Survivors included: his wife, Janey Holloway, of the home; Janey's children, Kate Nothstine (Chris), Barnard, Mo., Julia Brooke (Jeff), St. Joseph and Phillip Glick (Bernadette), Fenton, Mo.; Janey's grandchildren, Evan and Gabe Nothstine, Jacob and Ben Brooke and Catherine, Ashley and Phillip Glick; sisters, Anne Hegna (Peter), Minnesota and Kathy White (David), Wisconsin; sister-in-law, Judy Holloway; and, numerous nieces, nephews and other family members. He was buried in St. Joseph Memorial Park Cemetery.[78]

Pamela Marie Watson

Pam received her bachelor of science degree in Historic Preservation from Southeast Missouri State University in 2004. She completed her master's in Arts in Historic Preservation at the University of Maryland in May 2006. She accepted a position as a preservation planner with the City of Frederick, Maryland. Pam enjoyed the outdoors, 1000 Islands, the Adirondacks, and wood craftsmanship, especially creating her handmade kayaks.

Pam died at age 66 on October 23, 2015, in Skaneateles, New York. She was predeceased by her parents, Robert and Mary Quinn; sister, Nancy Burns; brother, Michael Quinn and nephew, David Burns. She was survived by her siblings, Patricia Quinn, Susan (John) Tartaglia, Colleen (Lee) Massenzio, Kathleen (Jason) Schoenborn and Sharon (Jeff) Carkner; brother-in-law, David Burns, several nieces, nephews, aunts, uncles and several cousins.[79]

CENTER

Endnotes

[1] 1982 Sagamore, page 93. Courtesy Special Collections and Archives, SEMO, Kent Library. Also, Dr. Art Mattingly's recollection that follows.

[2] U.S. News and World Report counted 15 around 2000. usnews.com/best-colleges/historic-preservation-major-3012. In 2021, there are 53 "Best Historic Preservation and Conservation Colleges in the U.S.," according to universities.com/programs/historic-preservation-and-conservation-degrees. PreservationDirectory.org lists 48 "Undergraduate and Certificate Programs in Historic Preservation." According to collegefactual.com: In 2019-2020, historic preservation was the 294th most popular major nationwide with 319 degrees awarded. This represents a 6.6% reduction in historic preservation degrees awarded over the prior year's total of 340. Approximately 48 colleges in the U.S. offer a historic preservation degree of some kind." (all viewed 14 Oct 2021).

[3] womenshistory.org/education-resources/biographies/ann-pamela-cunningham (viewed 27 May 2020).

[4] HP101 lecture notes, etc., might also be found in the personal papers of HP alumni, like David W. Jackson (BSHP, 1993), who have donated their original HP coursework to SEMO's Special Collections and Archives in Kent Library.

[5] momainstreet.org/cape-girardeau and facebook.com/oldtowncape (viewed 27 May 2020).

[6] "President's Report," *Southeast Missourian* (Cape Girardeau, Mo.), 19 Nov 1973.

[7] E-mail correspondence between Mattingly and David W. Jackson, 11 June 2020.

[8] Mattingly, Art. "Reflections on the Origins of the Historic Preservation Program," *Pastscript: News of the History Department,* Number 30 [actually, 31] (2006), 204; hereinafter, *Pastscript.* The transcript of the 2010 Mattingly interview only indicated that "Kelli" was the interviewer and "Jessica" was the videographer. The HP students were likely Jessica L. (Dickey) Balance (BSHP, 2012) and Kelli Mitchell (BSHP, 2015).

[9] The Center's website says 1978: semo.edu/history-anthropology/regionalhistory/index.html (viewed 26 Sept 2021).

[10] Bliss, Mark. "Southeast celebrating 25 years of teaching historic preservation," *Southeast Missourian* (Cape Girardeau, Mo.), 22 Apr 2006.

[11] dhe.mo.gov/cbhe/ (viewed 28 April 2018).

[12] Coordinating Board of Higher Education Minutes, May 10, 1980, describing action completed at the March 8, 1980, meeting of the CBHE Board. CBHE Board Books are archived online to 2003; prior to that, records are maintained in the Missouri State Archives. Pertinent copies of the 1980 agenda and minutes referencing SEMO's proposed HP Program, courtesy the Missouri State Archives, are filed in the SEMO HP Alumni Records Collection in SEMO's Special Collections and Archives.

[13] "Regional History Center Opens," and "Degree in Historic Preservation Approved," *Pastscript: News of the History Department*, Number 5 (1980-1981), 1.

[14] The Center's website says 1978. semo.edu/history-anthropology/regionalhistory/index.html (viewed 26 Sept 2021). Dr. Mattingly recollected in 2006 that it was 1979.

[15] *Pastscript,* Number 5, 1980-1981.

[16] Pastscript, 2009.

[17] *Pastscript*, 2009.

[18] semocatholic.org/history (viewed 26 Sep 2021). npgallery.nps.gov/GetAsset/20cbe283-700d-46d2-af12-2fbbd2794f0e; also posted at: mostateparks.com/sites/mostateparks/files/Wilson%2C%20J.%20Maple%20and%20Grace %20Senne%20House.pdf (viewed 26 Sept 2021).

[19] The Center's website says 1982: semo.edu/history-anthropology/regionalhistory/index.html (viewed 26 Sept 2021).

[20] *Pastscript*, Number 9 (1983-1984).

[21] Ramsey, B. Gene, "Chair Chat," *Pastscript*, Number 15 (1989-1990).

[22] Guide to the Historic Preservation Student Projects, 1970 (Series VII: Southeast Missouri State University, Box 1052, Folder 007): library.semo.edu/findingaids/Historic_Preservation.pdf (viewed 26 Sept 2021).
[23] semocatholic.org/history (viewed 26 Sept 2021).
[24] semocatholic.org/history (viewed 26 Sept 2021).
[25] semissourian.com/story/2423385.html (viewed 4 Oct 2021); also, at capecentralhigh.com/cape-photos/831-north-street/ Ken Steinhoff, blogger for Cape Girardeau History and Photos: Coming of Age in Cape Girardeau, provided the following news items from *The Missourian* about former occupants of 831 North St from 1941-1967:

December 24, 1941 – Among students home for the Missouri School of Mines at Rolla for the Christmas and New Year holidays are Alvin G. Haas, a senior in mechanical engineering, son of Mr. and Mrs. Alvin Haas, 821 Themis Street; Eddie Vandeven, a junior in electrical engineering, son of Mr. and Mrs. Edwin H. Vandeven, 831 North street; and John Leming, a freshman, son of Mr. and Mrs. Wm. E. Leming, 300 North Pacific street. [Editor's note: "William" was always spelled "Wm." in Missourian style. So far as I know, that's the only name that was routinely abbreviated.]

August 13, 1943 – At a meeting of a pinochle club with Mrs. Edwin H. Vandeven, 831 North street, Thursday afternoon, awards among the members were won by Mrs. John Knaup, Mrs. F.X. Roth and Mrs. Leslie J. Eagle, with Miss Lillian Nenninger receiving a guest prize. At an extra table of guests, including Mrs. John A. Vandeven, Mrs. John Nenninger, Mrs. Wm. Nenninger and Mrs. Joe Francis, awards went to Mrs. John Vandeven and Mrs. Wm. Nenninger.

October 27, 1947 – Mr. and Mrs. James Kearney, 831 North street, are the parents of a son, born at 3:50 o'clock Saturday afternoon at St. Francis Hospital. The child is the fourth in the family and fourth son. Mr. Kearney is associated with the Markham and Brown Construction Company.

December 28, 1948 – Eight Cape Girardeau residences, entrants in the annual outdoor Christmas lighting contest sponsored by the Retail Merchants Association will share in the $80 prize money following ratings by three out-of-town judges Monday night. For the section north of Broadway, first prize for the most artistic display was won by Mrs. James Kearney and sons, 831 North Street.

April 13, 1949 – Mr. and Mrs. A.P. Rueseler, 820 Merriwether street, have purchased the James Kearney home at 831 North street and expect to take occupancy of it within the next two or three weeks. Their daughter, Miss Ruth Rueseler, a student at the University of Missouri in Columbia, will be here to spend the summer vacation with her parents in their new home.

January 28, 1950 – In the midyear commencement class Wednesday at the University of Missouri wil be Miss Ruth Ruesler who is to be graduated with a B.S. degree in Elementary Education and who, on Feb. 6, will become a faculty member at Franklin School. She is the daughter of Mr. and Mrs. Albert P. Ruesler, 831 North street who will go to Columbia Tuesday to attend the ceremonies. The Ruselers now have with them for a weekend visit another daughter, Mrs. C. E. Burgett and her small daughter, Margaret Ruth, and young son, Mark Douglas, of Columbia.

August 28, 1950 – For Miss Ruth Rueseler, fiancee of Charles Knote of Cape Girardeau and daughter of Mr. and Mrs. A.P. Rueseler, 832 North street, Miss Rosemary Craig and her mother, Mrs. John Craig, honored the bride-elect Saturday afternoon at a crystal shower and dessert bridge.

February 16, 1951 – Numerous parties have been given in honor of Miss Ruth Gene Leming, fiancee of William Lewis Tucker of Jackson…. Mrs. A.P. Rueseler and her daughter, Mrs. Charles Knote, were hostesses at the Rueseler home, 831 North street, as friends greeted Miss Leming at a crystal shower.

December 30, 1967 – Miss Judy Carole Edwards and Ronald Charles Rueseler were united in marriage last Saturday afternoon at the Westminster Presbyterian Church in a double ring ceremony performed by the Rev. Thomas N. Bass. …The former Miss Edwards is the daughter of Mr. and Mrs. Harold Edwards of Kennett. Mr. Rueseler is the grandson of Mrs. Albert P. Rueseler, 831 North Street.

[26] news.semo.edu/hoffman-named-director-of-southeasts-bollinger-center-for-regional-history/ (viewed 27 Sept 2021).
[27] semissourian.com/story/2673217.html (viewed 15 Oct 2021).

[28] semo.edu/regional-history/index.html (viewed 14 Oct 2021).
[29] Architectural restoration was a fourth pillar; but is no longer a component. However, students do a lot of building documentation.
[30] Lucretia Drum, 56, chairwoman of the Cape County Democratic Party Central Committee, died Friday, April 26, 1997, at St. Francis Medical Center. Drum was known as an advocate for including more women in politics and helped organize President Bill Clinton's visit last year. She had heart trouble and was scheduled for heart surgery next week, friends said. Rick Althaus, county vice chairman, said, "I have a Ph.D. in political science, but I learned more about grass-roots politics from her than I did in my studies. From the time I began to be involved in the party in the 1980s, she essentially took me under her wing, she introduced me to people, she reminded me of details of local history and interpersonal dynamics." Funeral arrangements are incomplete with Cracraft-Miller Funeral Home in Jackson. https://www.semissourian.com/story/2700654.html?rand=4053 (viewed 20 Oct 2021).
[31] *HP Program Notes*, August 1996.
[32] *HP Program Notes*, December 1996.
[33] *HP Program Notes*, December 1999.
[34] oldmckendree.org/history (viewed 17 Oct 2021).
[35] rustedroutefarms.com; facebook.com/RustedRouteFarms; instagram.com/explore/locations/599321133560065/; and vrbo.com/2037403?fbclid=IwAR2zQA4eABuAYmyVsatnyB1pTqVJdX1hC2wgLQ8AdEy7GNGz2rr8QIMGcCI&noDates=true&unitId=2601923 (viewed 17 Oct 2021).
[36] Letter from Dr. Steven Hoffman to Mr. Donovan Rypkema, 8 Mar 2016. Rypkema was the keynote speaker at both the 25th and 35th SEMO HP Program anniversary banquets.
[37] Prepared by Carla Jordan; Kim Leazenby; Angela Moritz.
[38] Prepared by Tricia Bayer; Bill Baehr; Kirstin Glaser; Eric McNeil; Lori Ritchey.
[39] Prepared by Travis Ratermann; Catie Myers; Lauren Buchtmann; Sam Sampson; Rebecca Prendergast; Adrianna Salazar; Dawn Stricklin.
[40] Prepared by Adam Cashler; Elizabeth Ayers; Laura Stubblefield; Alan Higgins.
[41] Prepared by Muriel Anderson; Michael Bricknell; Edward Farrell; Megan Grisolano; Kali Morgan; Liz Murphy; Lauren Nobbe; Michael Painton; Alyssa Phares Jessie Rhodes; Amanda Russell; Sam Sampson; Sarah Stephens; Kate Warfield.
[42] Prepared by Matt Lingle; Hallie Fieser; Lauren Pollien; Megan Atchley; Sean Winstead; Marla Ives; Justin Gibbs; Karl Emch; Stephanie Palmer; Michael Painton; Wilson Newman.
[43] Prepared by Brenda Appleby; Kristen Lueders; Linda Brott; Seth McCoy; Joshua Doll; Megan Ritterbusch; Eric Kulengowski; Kate Warfield.
[44] Prepared by Lea Bohr; Samantha Bruer; Stephanie Chapman; Jessica Dickey; Allison Kelley; Cole Klawuhn; Amanda Lemon.
[45] Prepared by Heather Gurnow; Jessica Reisenbichler; Rebecca Schmitt; Stephanie Schwartz; Kelly Shaner.
[46] Prepared by Megan Atzert; Matt Caldwell; Cody Farrow; Chris Kinder; Heather Lamplough; Kelly Owens; Michael Parker; Jennifer Rowan.
[47] Costume Designer: Deana M. Luetkenhaus; Asst. Designers: Sarah Reeves; Kelcey Matheny; Erin Meadows; Stitchers: Whitney Buckner; Ashley Booker; Elizabeth Guebert; Kara Ratliff.
Asst. Designers: Sarah Reeves Kelcey Matheny Erin Meadows
Stitchers: Whitney Buckner Ashley Booker Elizabeth Guebert Kara
[48] Prepared by Vickie Chevalier; Allison Harvel; Deseray Helton; Samantha Kluesner; Kimberly Proctor.
[49] Prepared by Stephanie Buchanan; Matt Davis; Amy Gragert; Heather Hudnell; Sam Loos; Lindsey Lotz; Marla Moore.
[50] Prepared by Jessica Cox; Josh Bahr; Brandon Daake; Cheyenne French; Kayleigh Hamblin; Meghan Helm; Sawyer Magnus; Julie Nelson; Kim Trout; Riley Price; Adam Scholz; Cade Waddell; Kaylin White.

[51] Prepared by Scott Bates; Lucas Greenwalt; Samantha Harmon; Sarah La Venture; Maggie McDaniel; Kathleen Nelson; Stephanie Kerr; Nick Tamboli; Leah Redenour; Jacob Waters; Meighan Weaver.
[52] *Pastscript* 18, 1992-1993.
[53] *HP Program Notes*, January 1997.
[54] thejhchurchillfuneralhome.tributes.com/obituary/show/Kit-Wayne-Wesler-103165273 and constantgeography.com/2016/01/10/tribute-to-dr-kit-wesler-archaeologist-mentor-trekkie/ (viewed 4 Oct 2021).
[55] *HP Program Notes*, Spring 1997.
[56] News shared by Anne on Facebook in summer 2021. We are hoping Anne may donate the originals to the Special Collections and Archives Department in Kent Library.
[57] *Pastscript*, 1992-1993.
[58] *HP Program Notes*, August 1996.
[59] Per Dr. Steven Hoffman, in an e-mail exchange on October 16, 2021.
[60] Posted in 2010, with a directory of the HPA officers: app.semo.edu/cmpl/stuorgs/atoz/view_org_atoz.asp?org_id=109 (viewed 1 July 2021).
[61] www4.semo.edu/histpres/hpa/ (viewed 27 May 2018).
[62] Rypkema returned as keynote speaker at the 35th Anniversary Banquet, too.
[63] Invitation letter to David W. Jackson from Valerie Gaumont, President, Historic Preservation Association, December 18, 1995. Courtesy David W. Jackson Historic Preservation Program Collection, 1991-1993, SEMO Special Collections and Archives.
[64] Christy, David. "Campus News: Alum corrals preservation award," *Capaha Arrow*, 29 Mar 1995.
[65] capecentralhigh.com/cape-photos/southeast-missouri-state-university/arthur-mattingly-brought-history-to-life/ (viewed 3 Oct 2021).
[66] Obituary. Barbara Kaye Smith Hamblin, *The Banner-Independent* (Booneville, MS), 16 Apr 2020 (viewed 13 Oct 2021).
[67] *Pastscript,* 1991-1992.
[68] *Pastscript*, 1992-1993.
[69] *Pastscript*, 1992-1993.
[70] *Pastscript*, 1991-1992.
[71] Readers with more details, please contact the author and accept his sincere apologies for any errors or unintentional omissions. As above stated, this first attempt at such a compilation is published with acknowledged deficiencies.
[72] Nickell was an HP Program founder when he was chair of the Department of History, as previously recognized; taught folklore to nearly all HP students during his tenure; and, was a long time director of the Center for Regional History; along with a vitae too extensive to recount here.
[73] kellermanfoundation.org/about-us/ (viewed 28 Sept 2021).
[74] Cape Girardeau, Mo., 3 Oct 2017, SEMO news release: news.semo.edu/dr-arthur-h-mattingly-endowed-historic-preservation-fund-for-student-experiential-learning-established-at-southeast/ (viewed 3 Oct 2021).
[75] *Pastscript,* 2014.
[76] *Jefferson City News Tribune*, 12 Mar 2003; *Pastscript*, 2003.
[77] findagrave.com/memorial/212614828/lucas-barton-greenwalt (viewed 15 Oct 2021).
[78] *St. Joseph News Press*, 3 June 2012.
[79] *Pastscript*, 2007 and 2016; *The Post-Standard* (Syracuse, New York) 1 November 2015; edwardjryanandson.com.

Made in United States
Troutdale, OR
11/03/2026

56844373R00122